Sericulture

NIPA® GENX ELECTRONIC RESOURCES & SOLUTIONS P. LTD.
New Delhi-110 034

About the Reviewer and Author

Prof. Bhimasen Naik (b.1961) hails from an agri-horticultural family of Western Odisha (Village: Jamtalia, Tehsil: Sundargarh, District: Sundargarh). He obtained B.Sc. (Ag. & A.H.) and M.Sc. (Ag.) in Genetics and Plant Breeding from Chandra Shekhar Azad University of Agriculture and Technology, Kanpur in 1981 and 1983, respectively, and Ph.D. from Utkal University, Bhubaneswar in1999. He has been awarded ICAR Junior Research Fellowship and CSIR Senior Research Fellowship during his studies. He is a Fellow of the Indian Society of Genetics & Plant Breeding, New Delhi; and life member of the Crop Improvement Society of India, Ludhiana; the Indian Society of Plant Breeders, Coimbatore; the Society for Advancement of Rice Research, Hyderabad; the Indian Society of Oilseeds Research, Hyderabad and the Society for Plant Biochemistry & Biotechnology, New Delhi. He joined Odisha University of Agriculture and Technology (OUAT) as Research Assistant at Regional Research Station (later renamed Regional Research and Technology Transfer Station), Semiliguda, Koraput in 1984. Then he worked as Junior Breeder (Assistant Professor) at High Altitude Research Station, OUAT, Pottangi, Koraput (1984-1990); Assistant Professor and Assistant Professor (Senior Scale) at College of Agriculture, OUAT, Bhubaneswar (1990-1998); Training Associate (Horticulture)/Assistant Professor (Senior Scale) at Krishi Vigyan Kendra, Kendrapara (1998-1999); Junior Breeder/Assistant Professor (Senior Scale), Associate Professor and Professor at Regional Research and Technology Transfer Sub-Station (RRTTSS), OUAT, Jashipur, Mayurbhanj (1999-2010); Professor at Regional Research and Technology Transfer Station (RRTTS), OUAT, Chiplima, Sambalpur (2010-2021); Associate Director of Research, RRTTS, OUAT, Chiplima, Sambalpur (2010-2014); and Associate Dean, College of Horticulture, OUAT, Chiplima, Sambalpur (2016-2017). He retired from OUAT service in January 2021. He has worked on breeding and genetics of ginger, turmeric, mungbean, linseed and rice. He has more than 36 years of teaching, research and extension experience at Odisha University of Agriculture and Technology. He has 56 research papers published in peer-reviewed national and international journals, 23 abstracts of symposia, 10 books, 1 book chapter (Springer-Nature) and 20 popular articles to his credit. He has guided one M.Sc. (Ag.) and one Ph.D. student. He is associated in the development of six cultivars in ginger (Suruchi), turmeric (Surama, Ranga and Rashmi), linseed (Arpita) and heat tolerant rice (OUAT Kalinga Rice-8: Suryashree);

collaborated in the development of three cultivars in rice (Ashutosh, Gobinda and Hasanta) and registered one genetic stock of mungbean (BSN 1, INGR 00011). He is a reviewer of 'The Indian Journal of Agricultural Sciences' and 'Indian Farming' of the Indian Council of Agricultural Research, New Delhi.

Dr. Subhalaxmi Roy (b.1989) obtained her B.Sc. (Ag.) from College of Agriculture, Chiplima under Odisha University of Agriculture and Technology (OUAT) in 2012. She has received her M.Sc. (Ag.) in Entomology from BCKV, West Bengal and Ph.D. from Odisha University of Agriculture and Technology, Bhubaneswar in 2014 and 2021 respectively. She successfully qualified NET in Entomology in 2014 conducted by ASRB, New Delhi. She is currently working as Assistant Professor of Entomology at Institute of Agricultural Sciences, Siksha 'O' Anusandhan (Deemed to be University), Bhubaneswar since 2018. During her six years of teaching experience, she has taught 3 UG, 3 PG and 1 Ph.D. courses and still continuing. She has 18 research papers in journals of national and international reputes, 10 abstracts of seminars/symposiums/conferences, 2 books, 3 book chapters, 1 practical manual and 19 popular articles to her credit. She has attended 8 training programmes and 13 seminars/symposiums/conferences. She was awarded Young Researcher Award presented on 4th International conference held at Malla Reddy university, Hyderabad, in 2023. She has guided 4 M.Sc. (Entomology) students for their thesis. She is a life member of Society for Plant Protection and Environment, Society of Biotechnology and Bioinformatics and Indian society of Agriculture and Horticulture Research Development. She is also actively involved in various activities of the current working institution level.

Prof. Bijoy Kumar Mishra (b.1955) obtained B.Sc. (Ag.), M.Sc. (Ag.) and Ph.D in Entomology from Odisha University of Agriculture and Technology, Bhubaneswar in 1976 and 1978 and 1995 respectively. He started his carrier as Senior Research Assistant in OUAT in 1980. Subsequently he worked as Subject Matter Specialist, Assistant Professor, Senior Scientist, Professor (Entomology), Head of the Department (Entomology) and Dean, College of Agriculture, OUAT before superannuating in 2015. Dr. Mishra is having 35 years of experience in teaching, research and extension. He has guided 18 PG and 3 Ph.D. students and has published 112 research papers in journals of national and international repute with 4 practical manual, 2 books and 3 book chapters. He has handled many Govt. Funded projects during his research tenure. He has attended many summer and winter schools and training programme in biological control field during his teaching period. Besides being a reputed scientist in biological control of crop pests, Dr. Mishra is also having vast knowledge in the field of beneficial insects, host-plant resistance and IPM of crop pests. He is a life member of several societies in the field of Entomology. During his period of teaching, he was involved in various college level activities. He has been selected as an external question setter for the students of B.Sc. (Ag.) and M.Sc. (Ag.) for different SAUs. He has also delivered many radio talks pertaining to his field of specialization benefitting to farmer's interest. Presently he is working as Professor and Head in Entomology, Faculty of Agricultural Sciences, SOA (Deemed to be university), Bhubaneswar.

Sericulture Skill Enhancement

Reviewer

Bhimasen Naik

Former Associate Dean
College of Horticulture, OUAT
Chiplima, Sambalpur, Odisha, India

Author

Subhalaxmi Roy

Assistant Professor of Entomology
Institute of Agricultural Sciences,
Siksha 'O' Anusandhan Deemed University,
Bhubaneswar, Odisha, India

Bijoy Kumar Mishra

Former Professor and Head of Entomology
Institute of Agricultural Sciences,
Siksha 'O' Anusandhan Deemed University,
Bhubaneswar, Odisha, India

NIPA® GENX ELECTRONIC RESOURCES & SOLUTIONS P. LTD.
New Delhi-110 034

NIPA® GENX ELECTRONIC
RESOURCES & SOLUTIONS P. LTD.

101,103, Vikas Surya Plaza, CU Block
L.S.C.Market, Pitam Pura, New Delhi-110 034
Ph : +91 11 4386 0225, 9717133558, 9540816132
E-mail: newindiapublishingagency@gmail.com
Website: www.niparesources.com

Print ISBN: 978-93-58871-95-1

ebook ISBN: 978-93-58877-57-1

NIPA® also publishes books in a variety of electronic formats. Some content that appears in print may not be available in electronic books, and vice versa

Composed and Designed by NIPA®.

"The book is good when it says only what should be said."
- Aristotle

DEDICATED
TO
OUR MOTHERS

Prof. (Dr.) Hari Har Ram
Former Professor and Head, Vegetable Science, GBPUAT, Pantnagar
Former Vice President and Chief R&D, Krishidhan Veg. Seeds, Pune
Former FAO Consultant (Myanmar, Libya, and DPR Korea)

Foreword

The present book entitled "Sericulture: Skill Enhancement" reviewed by Prof. Bhimasen Naik, Former Associate Dean, College of Horticulture, OUAT, Chiplima, Sambalpur, Odisha, India, and authored by Dr. Subhalaxmi Roy, Assistant Professor, Entomology, Institute of Agricultural Sciences, Siksha 'O' Anusandhan Deemed University, Bhubaneswar, Odisha, India, and Prof. Bijoy Kumar Mishra, Professor and Head, Entomology, Institute of Agricultural Sciences, Siksha 'O' Anusandhan Deemed University, Bhubaneswar, Odisha, India, is as per recommendations of sixth Deans' Committee of Indian Council of Agricultural Research, New Delhi on revision of courses and their outlines for UG students to be followed by all ICAR institutes, state agricultural universities and other colleges uniformly throughout the country. I could know Prof. Naik through several email and WhatsApp exchanges as we both happened to be authors of several books for a common publisher, New India Publishing Agency, New Delhi. He gave me opportunities to read some of his manuscripts and that is where I could realise his passion for teaching and writing text books as per exact needs of the students in accordance with the prescribed course outline so that the course teacher and the students could get the entire needful information in most concise manner at one place and in pursuit of this goal, he could write text books which in a way reflect the mirror image of an ideal note book for a particular course.

Following up his passion for text book writing and as requested by New India Publishing Agency, New Delhi, he entered into dialogue with New India Publishing Agency, New Delhi and accepted the challenge of arduous task producing manuscripts for all the courses related to Horticulture for the undergraduate courses exactly as per requirements of the recently revised syllabus of B.Sc. (Hons.) Horticulture by ICAR which is uniform throughout the country. He knew that he himself and or supported by few co-authors cannot write all the needed text books covering entire Horticulture as applicable to the UG students and that too in a time bound manner. Therefore, he initiated dialogue and discussion among his former students, now occupying respectable teaching/research positions in the universities and professional colleagues and prepared list of authors for each course. For some of the courses, he offered himself as main author or supported by co-authors. However, for the entire series of the books under this project numbering 77 just to ensure uniformity regarding headings, subheadings, style of writing, simple language to be easily understood by students, formulation of small questions at the end of each chapter, inclusion of relevant updated information and references and above all to make the entire manuscript simply flawless, he offered his services as reviewer of each manuscript under this series of the text books. In a way, he took the entire responsibility of producing manuscripts of all the 77 books under this series in accordance with recently revised syllabus of B.Sc. (Hons.) Horticulture which is uniform throughout the country, with utmost care, dedication, sincerity and was not prepared to accept any lame excuses for anything going haywire. I have gone through Prof. Naik's biography and I am fully convinced that sincerity, honesty, dedication, finishing given job on time and selfless hardworking attitudes and disciplined way of working are hallmark of his personality and adhering to these qualities, make him different from the masses. He has galvanized his team of authors and coauthors and has kept a close watch on the progress of each manuscript to ensure that each manuscript is delivered to the publisher as per timeline given in the contract. Full appreciations and congratulations to Prof. Naik.

The present book is the outcome of Prof. Naik's mammoth project on text book writing as per prescribed syllabus by ICAR for B.Sc. (Honours) Horticulture and strictly adhering to a time bound manner approach. There are some Skill Enhancement Courses (SEC) in this recently revised new syllabus, each of 2 (0+2) credits, aiming at developing skill. Sericulture is one of such courses. The present textbook "Sericulture: Skill Enhancemen" is designed to be used in practical classes of one semester. It has 20 chapters for 32 practical classes. The authors of this book, namely, Dr. Subhalaxmi Roy, Assistant Professor, Entomology, Institute of Agricultural Sciences, Siksha 'O' Anusandhan Deemed University, Bhubaneswar, Odisha, India, and Prof. Bijoy Kumar Mishra, Professor and Head, Entomology, Institute of Agricultural Sciences, Siksha 'O' Anusandhan Deemed University, Bhubaneswar, Odisha, are well known in their subject and have delivered praiseworthy text book on sericulture for B.Sc. (Honours) Horticulture under supervision and constant encouragement of Prof. Naik, the reviewer of the book. The glossary and the list of figures are the added desirable features of the book. I have no hesitation in congratulating Prof. Naik, Dr. Subhalaxami Roy and Prof. Bijoy Kumar Mishra for preparing this text book for UG students of Horticulture and I am sure students will find this book as per their liking as the book is precise, concise and is in note-book style. I fully endorse the sentiments of the reviewer and the authors about this book who in the beginning of this book have quoted Aristotle "The book is good when it says only what should be said." Yes, the book says what should be said.

Hari Har Ram

Preface

The Indian Council of Agricultural Research has recently revised the syllabus of B.Sc. (Hons.) Horticulture which is uniform throughout the country. There are some Skill Enhancement Courses (SEC) in the syllabus, each of 2 (0+2) credits, aiming at developing skill. Sericulture is one of them. The present textbook "Sericulture: Skill Enhancement" is designed to be used in practical classes of one semester. It has 20 chapters for 32 practical classes.

Simple and lucid language has been followed for easy understanding of the beginners. The information contained in the textbook has been gathered from various published sources and internet websites which are mentioned at the end of each chapter under references. Attempts have been made to provide latest information; still some valuable information might have been missed. Questions are set at the end of each chapter under outcomes assessment to assess the understanding of the students. The questions include true/false questions, one/two sentence answer questions, short notes, and multiple-choice questions (MCQs).

We would like to thank Mr. Sai Suman, 3rd year B.Sc. (Ag) for his immensible efforts in helping in preparation of the book in stipulated time period.

We have tried our best to remove the errors, typographical or otherwise, from the text; still there might be some. We would highly appreciate if it is brought to our notice for rectification in next edition. We cherish the encouragement and cooperation received from our family members during preparation of the manuscript. We congratulate M/S NIPA® GENX Electronic Resources & Solutions P. LTD., New Delhi for their support and publishing in a short time.

Though the textbook is primarily intended for B.Sc. (Hons.) Horticulture students, the counterparts of B.Sc. (Hons.) Agriculture also may be benefitted.

We seek suggestions from learned teachers and students for further improvement of the textbook.

Kuchinda, Sambalpur, Odisha — **Bhimasen Naik**
Bhubaneswar, Odisha — **Subhalaxmi Roy**
Bhubanesear, Odisha — **Bijoy Kumar Mishra**

Contents

Glossary

A

Accessory Gland – In silkworms, accessory glands are part of the reproductive system that secrete substances essential for egg fertilization and cocoon formation.

Aided Grainage – A controlled method of silkworm egg production where environmental conditions like temperature and humidity are regulated to ensure healthy egg development and hatching.

All Silk – A term used to describe fabrics or products made entirely from pure natural silk without blending with synthetic or other fibres.

Ampatia – A disease in silkworms caused by pathogens like bacteria or viruses, leading to abnormal development and affecting cocoon production.

Anaphe Silk – A type of wild silk obtained from Anaphe moth species, mainly found in Africa, known for its soft texture and natural elasticity.

Anterior Silk Gland – One of the three silk-producing glands in silkworms (along with the middle and posterior silk glands) that plays a role in converting liquid silk protein into solid silk fibres.

Antiseptic – Chemical agents used in sericulture to disinfect rearing trays, silkworm eggs, and rearing houses to prevent diseases and ensure healthy silkworm growth.

Arboriculture – The cultivation and management of trees, particularly those like mulberry trees, which are grown for feeding silkworms in sericulture.

Art Silk (Artificial Silk) – A synthetic fibre that mimics the appearance and texture of natural silk, often made from rayon, polyester, or other artificial materials.

Automatic Silk Reeling Machine – A mechanized device used to unwind silk fibres from cocoons, improving efficiency and quality in silk production compared to traditional manual reeling methods.

B

Bale – A bundle of raw silk or silk cocoons that are compressed and packed for storage or transportation.

Bed Cleaning – The process of removing leftover mulberry leaves, silkworm litter (excreta), and dead larvae from silkworm-rearing trays to maintain hygiene and prevent disease.

Basic Irrigation – The fundamental watering process used for mulberry plantations to ensure optimal growth and leaf production, which is essential for silkworm feeding.

Biotype – A group of silkworm populations that share similar genetic traits and characteristics, which can influence silk quality and productivity.

Bivoltine – A type of silkworm breed that produces two generations per year, commonly found in temperate regions and known for producing high-quality silk.

Bleaching Powder – A disinfectant used in silkworm rearing houses.

Brood - A batch of silkworm eggs or larvae reared together.

Brushing - Transferring newly hatched silkworm larvae to rearing trays.

C

Castor – A type of plant (*Ricinus communis)* whose leaves are used as food for eri silkworms *(Samia ricini*), which produce Eri silk.

Chamber Steaming – A process where silkworm cocoons are subjected to steam in a closed chamber to kill the pupae inside, preventing them from breaking the silk filament during emergence.

Chandrika – A local term in some regions of India referring to a compact silk cocoon of good quality with high silk content.

Charkha – A traditional hand-operated silk-reeling device used to unwind silk fibres from cocoons and convert them into raw silk threads.

Chawki – The early-stage (first and second instar) silkworm larvae that require special care, including controlled temperature, humidity, and high-quality mulberry leaves for better survival and growth.

Cocoon – The protective covering made of continuous silk filament that silkworms spin around themselves during the pupal stage before transforming into moths. It is the raw material for silk production.

Cocoon Stuffing – A process in which defective or cut cocoons are filled with cotton or other materials to be used for decorative or commercial purposes.

Cocoon Spinning – The stage where mature silkworms extrude silk filament through their spinnerets and wrap themselves in it, forming a cocoon.

Cold Reeling – A silk reeling technique where silk filaments are extracted from cocoons in cold water instead of hot water, typically used in certain traditional silk-making processes.

Compact Cocoon – A high-quality cocoon with tightly packed silk layers, which yields a higher percentage of silk during reeling.

Copulation – The mating process of male and female silkworm moths, which results in the fertilization of eggs for the next generation of silkworms.

Cut Cocoon – A cocoon that has been cut open to extract the pupa, often used for seed production (breeding) or in crafts and other non-textile applications.

D

Damping Off – A fungal disease affecting young silkworm larvae or mulberry seedlings, causing wilting, decay, and high mortality rates in early stages.

Denier – A unit of measurement for the thickness (fineness) of silk filament, defined as the weight in grams of 9,000 metres of silk fibre. A lower denier indicates finer silk.

Diapause – A period of suspended development in silkworm eggs, typically induced by environmental factors such as temperature and humidity, allowing synchronization of rearing cycles.

Disease-Free Laying (DFL) – A batch of silkworm eggs that have been scientifically tested and certified as free from diseases, ensuring better hatchability and healthy larvae.

Domestication – The process of breeding and rearing silkworms in controlled environments over generations for improved silk production, making them dependent on human care.

Double Cocoon – A cocoon formed when two silkworm larvae spin silk together, creating an irregular structure unsuitable for reeling continuous silk filament but used for spun silk production.

Downy Mildew – A fungal disease that affects mulberry leaves, causing yellowish patches, curling, and reduced leaf quality, ultimately impacting silkworm nutrition and silk yield.

Dry Reeling – A method of reeling silk from cocoons without immersing them in hot water, often used for certain types of wild silk like Tussar, where sericin is retained for texture.

Dupion – A type of silk thread produced when two silkworms spin their cocoons together, resulting in a slightly irregular, textured silk fabric often used in high-end textiles.

E

Early Moth Emergence – A condition where silkworm moths emerge from their cocoons earlier than expected, often due to temperature fluctuations or improper storage conditions, which can affect breeding cycles.

Ecdysis – The process of moulting in silkworms, where they shed their old skin to grow during different instar stages. This occurs multiple times before the silkworm reaches maturity.

Egg Hatching – The process in which silkworm eggs hatch into larvae (chawki worms) under optimal temperature (24–26°C) and humidity (75–85%) conditions.

Egg Washing – A process used in silkworm egg production to remove harmful pathogens and improve hatchability, typically done with disinfectants like formalin.

Eri Silk – A type of silk produced by the *Samia ricini* silkworm, which feeds on castor leaves. Unlike other silks, Eri cocoons are open-ended, allowing the silk to be spun rather than reeled.

Exuviae – The shed skin of silkworms after moulting (ecdysis), which remains in the rearing trays and must be cleaned out to maintain hygiene.

F

Fecundity – The reproductive capacity of a silkworm moth, measured by the number of eggs laid by a female moth. High fecundity ensures better silkworm seed production.

Fibre – The silk filament spun by silkworms to form cocoons, composed mainly of fibroin and sericin proteins, which are later processed into raw silk threads.

Fibroin – The core structural protein of silk fibres, responsible for the strength and durability of silk. It makes up about 70–80% of the silk filament.

Filippi's Gland – A gland in silkworms that secretes a fluid aiding in the attachment of silk threads during cocoon spinning.

Flacherie – A bacterial or viral disease in silkworms that causes sluggishness, vomiting, and liquefaction of the body, leading to high mortality and reduced silk yield.

Formalin – A chemical disinfectant (aqueous solution of formaldehyde) used in sericulture to sterilize rearing trays, equipment, and rearing houses to prevent silkworm diseases.

Foss Silk – A defective silk filament with uneven thickness and poor quality, often caused by irregular cocoon spinning or improper reeling conditions.

Fumigation – A method of disinfecting silkworm-rearing houses, egg chambers, and grainages using smoke or chemical vapours (e.g., formalin) to eliminate pathogens and pests.

G

Gestation Period – In sericulture, it refers to the developmental time between the fertilization of silkworm eggs and their hatching.

Grainage – A facility or process where silkworm eggs are produced, stored, and treated for disease prevention before distribution to farmers.

Grasserie – A viral disease in silkworms caused by a Baculovirus, leading to swelling, translucent skin, and white granular deposits inside the body. It reduces cocoon quality and production.

Green Cocoon – A cocoon that appears greenish due to its natural pigmentation, common in certain silkworm breeds like Muga silk-producing worms.

H

Harvesting of Cocoon – The process of collecting mature silkworm cocoons from mountages after the spinning phase is complete.

Harvesting of Leaves – The collection of mulberry or other host plant leaves at the right maturity stage to feed silkworms

Hatching – The emergence of young silkworm larvae from fertilized eggs, which requires controlled temperature and humidity conditions.

Herold's Gland – A gland in silkworms that secretes a lubricating fluid to help in the smooth movement of silk filament during cocoon spinning.

Hibernation – A resting phase in silkworm eggs (diapause stage) during winter to synchronize their life cycle with favourable rearing seasons.

I

Improved Charkha – A modified version of the traditional hand-operated silk reeling machine, designed to improve efficiency and silk quality.

Inbreeding – The practice of mating silkworms from the same genetic line, which can lead to reduced vigour and productivity due to genetic weaknesses.

Ishiwata Gland – A specialized gland in silkworm larvae that aids in silk fibre formation, named after the scientist who discovered it.

J

Jali – A net or perforated sheet used in silkworm rearing to facilitate aeration, easy cleaning, and cocoon spinning.

Juvenile – The young stage of the silkworm before it reaches maturity, primarily referring to the larval stages.

L

Late Age Rearing – The rearing of silkworms in their third, fourth, and fifth instar stages, requiring larger quantities of leaves and higher care for better cocoon yield.

M

Malformed Cocoon – A defective cocoon that is irregular in shape or structure, often caused by diseases, genetic issues, or poor rearing conditions.

Milky Disease – A bacterial disease in silkworms that causes a milky-white appearance in their body fluids, leading to high mortality and loss in silk production.

Moriculture – The scientific practice of cultivating mulberry plants for feeding silkworms, involving improved agronomic techniques for better leaf yield.

Mother Moth – A female silkworm moth selected for egg production after ensuring disease-free status.

Mother Moth Examination – The process of testing and screening female moths for diseases (especially pebrine) before using their eggs for seed production.

Moulting – The periodic shedding of the old skin by silkworm larvae as they grow through different instar stages.

Mountage – A structure or device (like bamboo trays or plastic mountages) where mature silk-worm larvae are placed for cocoon spinning.

Muga – A rare and high-quality golden-yellow silk produced by the *Antheraea assamensis* silkworm, native to Assam, India.

Multivoltine – A silkworm breed that produces multiple generations in a year, typically found in tropical climates, such as certain varieties of *Bombyx mori.*

N

Non-Mulberry – Silk varieties obtained from silkworms that do not feed on mulberry leaves, such as Eri, Tasar, and Muga silk.

O

Oak Tasar – A variety of wild silk produced by *Antheraea proylei* silkworms that feed on oak leaves, mainly found in Himalayan regions.

Over Winter – The process of preserving silkworm eggs in cold conditions to induce diapause, ensuring synchronized hatching in the next season.

P

Pebrine – A highly contagious silkworm disease caused by the microsporidian parasite *Nosema bombycis,* leading to poor cocoon formation and high mortality.

Pruning – The practice of cutting mulberry branches to promote fresh leaf growth, ensuring high-quality food for silkworms.

R

Race – A specific strain or breed of silkworm, developed based on characteristics like voltinism (univoltine, bivoltine, multivoltine) and silk yield.

Raw Silk – The silk fibre obtained from cocoons through reeling, before any further processing like dyeing or weaving.

Rearing – The process of growing silkworms from the larval stage to the cocoon-spinning stage by providing suitable environmental conditions and feeding.

Reelability – The ability of a cocoon to be reeled efficiently, determined by factors such as cocoon quality, filament length, and silk yield.

Ring Spinning – A modern spinning technique used to twist silk fibres into a stronger and finer yarn, improving the quality of silk threads.

Ring Tail – A defect in silkworms where the larva's tail remains curled due to improper growth or disease, leading to cocoon abnormalities.

S

Sanitation – Maintaining cleanliness in silkworm-rearing houses, grainages, and reeling units to prevent diseases and ensure healthy silkworm development.

Spun Silk – Silk yarn made from broken or cut silk filaments, usually obtained from defective or pierced cocoons. It is softer but less lustrous than reeled silk.

Stadium – The period between two successive moults in a silkworm's life cycle (e.g., the first instar stadium is the time between hatching and the first moult).

Staining – The discolouration of cocoons caused by bacterial or fungal infections, improper handling, or dead pupae, reducing silk quality.

Stifled Cocoon – A cocoon in which the pupa inside has been killed using heat or steam to prevent the emergence of the moth and preserve the silk filament for reeling.

Stiffing – A process of drying cocoons to kill the pupae, usually done through sun drying, hot air, or steaming, to maintain the silk quality.

T

Tapping – A traditional method of testing the maturity of silkworms before cocoon formation by gently tapping their bodies to check for softness and readiness to spin.

Tasar – A type of wild silk produced by the *Antheraea mylitta* silkworm, which feeds on Arjun and Asan trees. It is known for its rich texture and durability.

Tensile Strength – The strength of silk fibres measured by their ability to withstand tension without breaking, an important quality parameter for silk threads.

Thrown Silk – Silk thread that has been twisted or doubled to improve strength and uniformity, often used for weaving fine silk fabrics.

Toxicity – Harmful effects caused by pesticides, disinfectants, or pathogens on silkworms, leading to illness, mortality, and reduced silk production.

Tran Silk – A term used for transitional silk, which refers to silk obtained from cocoons that have started to deteriorate but are still usable for lower-grade silk products.

Tukra Disease – A viral disease in mulberry plants caused by Maclura virus, leading to leaf curling, yellowing, and reduced leaf yield, affecting silkworm nutrition.

U

Univoltine – A type of silkworm breed that produces only one generation per year, typically found in temperate climates, producing high-quality silk.

Uzi Fly – A serious parasitoid pest (*Exorista bombycis*) that lays eggs on silkworm larvae; the maggots feed inside, killing the worms and reducing cocoon production.

List of Figures

1

Introduction to Sericulture

1.1 Introduction

Sericulture involves raising silkworms to produce silk. It is an agricultural industry that includes growing food plants for silkworms, raising them to produce cocoons, and then processing the cocoons to create yarn for weaving, resulting in value-added benefits.

1.2 Definition of Sericulture

Sericulture (also known as 'silk production') involves breeding and caring for silk worms to produce silk on a large scale. This process includes raising the worms on mulberry plants, harvesting and processing their cocoons to extract raw silk fibres, and ultimately creating commercial silk products. The art of silk production is called sericulture that comprises cultivation of mulberry, silkworm rearing and post-cocoon activities leading to production of silk yarn (FAO). While there are many species of silk worms, only a few are commonly used for commercial purposes.

1.3 Silkworm

The silkworm is the larval stage of the domesticated silk moth *Bombyx mori*, which is in latin also referred to as the "silkworm of the mulberry tree".

Being the main source of silk, it is a significant insect from an economic standpoint. White mulberry leaves are the primary meal of silkworms (monophagous). The generation of domestic silk moths is largely dependent on humans due to millennia of selective breeding. Because wild silk moths have not undergone selective breeding like their domestic cousins, they are not economically viable to produce silk.

The Neolithic Age was the beginning of the development of the tools needed to enable the production of bigger amounts of silk thread. It is still possible for the wild *B. mandarina* and the domesticated *B. mori* to breed and occasionally create hybrids.

1.4 Types of Silkworms

Five main commercially significant varieties of silk are produced by several kinds of silkworms which eat a variety of food plants: Other types of silks, outside mulberry, are commonly referred to as non-mulberry silks. All of these commercial types of silk are produced in India, which gives it a particular advantage.

1.4.1 Mulberry Silk

This variety produces the majority of the world's commercial silk, and silk is frequently used to refer to mulberry silk. ***Bombyx mori***, a silkworm that only consumes the leaves of the mulberry plant, is the source of mulberry silk (**Fig. 1.1**). These silkworms are raised indoors and are fully domesticated. The five states that produce the most mulberry silk in India are Karnataka, Andhra Pradesh, West Bengal, Tamil Nadu, and Jammu & Kashmir. Combinedly these states produce 92% of the nation's mulberry raw silk.

Fig. 1.1: Mulberry silkworm (*Bombyx mori*)

1.4.2 Tasar Silk

Tasar (Tussah) is a coarse silk with a coppery hue that is mostly utilised for interior design and furniture. It has a different feel and appeal from mulberry silk, while being less glossy. The silkworm ***Antheraea mylitta***, which mostly feeds on the food plants Arjun and Asan, produces tasar silk. The rearings take place outside on exposed trees. In addition to Maharashtra, West Bengal and Andhra Pradesh, the states of Jharkhand, Chattisgarh and Odisha are the primary producers of tasar silk in India. The primary source of identity for many Indian tribal communities is Tasar culture.

Fig. 1.2: Tasar silkworm (*Antheraea mylitta*)

1.4.3 Oak Tasar Silk

This is a more refined type of tasar produced by the Indian silkworm, ***Antheraea proyeli***, which feeds on oak trees that grow naturally in the sub-Himalayan region of India, which includes the states of Jammu & Kashmir, Assam, Manipur, Himachal Pradesh, Uttar Pradesh and Meghalaya (**Fig. 1.3**). ***Antheraea pernyi***, a different type of silkworm, is the source of the oak tasar, which is primarily produced in China (**Fig. 1.4**).

Fig. 1.3: Oak tasar silkworm (*Antheraea proyeli*)

Fig. 1.4: Oak tasar silkworm (*Antheraea pernyi*)

1.4.4 Eri Silk

Eri, often referred to as Endi or Erandi, is a multivoltine silk that, in contrast to other types of silk, is spun from open-ended cocoons. The domesticated silkworm ***Philosamia ricini***, which mostly feeds on castor plants, produces eri silk (**Fig. 1.5**). Ericulture is a domestic practice mostly used to produce pupae rich in protein, which are a treat for the native people. As a result, the eri cocoons are spun and have an open mouth. These tribes people prepare chaddars, or wraps using silk that they make themselves. This culture is primarily prevalent in Assam and the northeastern states of India. Additionally, Bihar, West Bengal and Odisha have it.

Fig. 1.5: Eri silkworm (*Philosamia ricini*)

1.4. 5 Muga Silk

Fig. 1.6: Muga silkworm (*Antheraea assamensis*)

The pride of the Assamese state and a symbol of India's superiority is this golden yellow silk. ***Antheraea assamensis***, a semi-domesticated multivoltine silkworm, is the source of it. These silkworms are raised on trees resembling tasars and fed on the fragrant leaves of Som and Soalu plants. Assamese heritage and culture are deeply rooted in the Muga culture, which is unique to that state. A valuable product, muga silk is used to make items like *chaddars*, *mekhalas* and *sarees*.

1.5 History of Sericulture

Chinese sources indicate that silk manufacture from *B. mori* was discovered circa 2700 BC. Emperor Huang-ti is said to have asked empress Si-lung-Chi to investigate the reason behind the damaged mulberry leaves on their garden's trees. The empress discovered white worms consuming the foliage. She also noticed that they had gleaming cocoons surrounding them. A cocoon fell into her tea cup, and delicate strands detached themselves from the cocoon. The origin of the silk business was kept a secret for over 2,000 years in China. After a while, China's monopoly on the manufacture of silk was broken, and sericulture spread to Japan via Korea before reaching other nations. In India, sericulture has been expanding as an agro-based sector that is essential to the development of the rural economy.

The silk industry was saved in 1870 by the renowned French scientist Louis Pasteur, who demonstrated that the epidemic Pebrine sickness of silkworms could be prevented by performing a straight forward microscopic study on mature female moths. These developments paved the way for a more scientific and mechanised method of producing silk than had been used in the past.

Chinese silk was brought to India by traders, primarily from Samarkand and Bukhara, where it quickly became quite fashionable among the nobility and kings. Because royal families and temples in India demanded rich materials, Jamawar and other brocade weaving centres grew up in holy cities and commerce hubs. The ancient centres were primarily found in South India, Gujarat and Malwa. The major hubs for brocade weaving in the North were Delhi, Lahore, Agra, Fatehpur Sikri, Varanasi, Mau, Azamgarh and

Murshidabad. The most well-known locations for Jamawar weaving were in South India, Gujarat, Assam and Malwa.

1.6 Present Status of Sericulture in India and Around World

India ranks as the world's second largest producer of silk. About 9.2 million individuals in rural and semi-urban areas of the nation are employed in the sericulture sector (as per 2022-23). This sector ranks among the top earners of foreign exchange in the nation. The 52,360 villages in India are home to sericulture operations. Four varieties of natural silks are produced in India: Muga, Eri, Tasar, and Mulberry. Using the raw material, the nation creates silk clothing, accessories, made-ups, carpets, shawls, scarves, cushion coverings and yarns.

36,582 metric tonnes (MT) of silk were produced in India in 2022-2023. In 2021-2022, India's total silk production was 34,903 MT, up 3.4% year-over-year from 33,770 MT of 2020-2021. Among the various forms of silk produced in the nation, mulberry production makes up the largest portion. The states that produce the most silk in the nation include Tamil Nadu, Uttar Pradesh, Assam, Bihar, Gujarat, Jammu & Kashmir, Karnataka, Chhattisgarh, and West Bengal. Approximately 32.3% of the nation's total silk production in 2022-2023 came from Karnataka. The state of Andhra Pradesh came next, accounting for 16.0% of the nation's total silk production in 2022-2023.

The two largest producers of silk are China and India, and the industry is still very important to the world economy. With a yearly production of over 175,000 metric tonnes, China has the largest share of the market, generating almost 80% of all silk produced worldwide. India is the world's second-largest producer, with an annual production of over 36,000 metric tonnes, or almost 16% of the total. With a compound annual growth rate (CAGR) of 8.4%, the worldwide silk market is predicted to rise rapidly, reaching a value of $20 billion by 2024 and then reaching $44.6 billion by 2034. The textile sector, which continues to rule the market because silk is associated with luxury and high-quality fashion, is the driving force behind this rise. Significant silk-producing nations also include Brazil and Uzbekistan. Brazil produces about 700 metric tonnes annually, with year-round production aided by its tropical climate, whereas Uzbekistan produces roughly 1,000 metric tonnes annually, concentrating on mulberry silk.

1.7 Scope of Sericulture

Sericulture is an agriculturally based, lucrative business that involves a variety of activities and has the ability to provide a large number of jobs. It is a major factor in determining the economic future of rural communities.

The long series of interconnected processes that go into making silk fabric—such as mulberry cultivation, silkworm breeding, silk production, twisting, warp and weft creation, dying and printing, spun silk production, finishing silk fabric, designing silk clothing, etc.—provide rural and semi-urban people with a means of subsistence. The availability of human resources and India's year-round suitable climate have been the country's greatest advantages in the silk business. Sericulture creates jobs all year round and offers substantial returns with little initial investment and a transferable return in a short amount of time. It is clear that sericulture, by creating jobs, revenue, and foreign exchange, is essential to the nation's economic growth.

1.8 Importance of Sericulture

Sericulture, the process of raising silkworms for textile production, is significant for a number of reasons. Sericulture, especially in China, India and Thailand, is a major source of income for many rural families. It helps to reduce poverty and promote rural development by giving millions of people—especially women and small farmers—a job. Due to its high value and robust demand in the international market, silk generates sizable export profits. Mulberry tree cultivation is encouraged by sericulture, and when combined with other agricultural pursuits, it can enhance production and land usage. Compared to other farming methods, it has a comparatively low initial investment requirement, which makes it affordable for small and marginal farmers. Because sericulture promotes sustainable agriculture and uses few chemicals and pesticides, it is environmentally safe. Mulberry leaves, silkworm pupae, and the protein byproduct sericin are examples of sericulture byproducts with a variety of applications in medicine, cosmetics and agriculture. Silk manufacture has historically and culturally been important in many civilizations, supporting traditional crafts and legacy. Sericulture helps farmers and craftspeople improve their technical knowledge and skill sets which increases their chances of making a living. The industry promotes technological development and scientific research to raise the standard and efficiency of silk production. Mulberry tree cultivation for sericulture contributes to ecological balance and biodiversity conservation.

1.9 Problems of Indian Sericultural Industries

Despite making a substantial economic and cultural legacy contribution, the Indian sericulture business confronts a number of obstacles:

1. **Technological Gaps:** A lot of traditional sericulture methods are still used in India. To improve productivity and quality, modern technologies are required for the manufacturing of cocoons, reeling and weaving.

2. **Management of Pests and Diseases:** Sericulture is extremely vulnerable to pests and diseases. The health and productivity of silkworms are greatly impacted by insect infestations like yellow moths and disease outbreaks like pebrine and grasserie.

3. **Silk Quality:** Indian silk, particularly mulberry silk, is not always of constant quality. This has an impact on its marketability on national and worldwide levels. Improving sericulture methods and breeding procedures could improve the quality of silk.

4. **Infrastructure and Supply Chain:** The supply chain needed for the storage, transportation and processing of cocoons is insufficient, and the supply chain in sericulture is disjointed. High post-harvest losses and inefficiency are the outcome of this.

5. **Market Fluctuations:** Farmers' livelihoods are impacted by changes in silk cocoon prices. Contributing factors to these problems include market dynamics, inadequate pricing procedures and restricted market access.

6. **Financial Constraints:** Formal credit and financial services are not readily available to many sericulturists. Their capacity to make investments in improved technology, high-quality seeds and disease control strategies is hampered by this.

7. **Skill Gap and Training:** Among sericulturists, there is a sizable skill gap. Insufficient training and extension services result in low productivity and less-than-ideal farming methods.

1.10 Geographical Indications of Indian Silk

Baluchari Saree - West Bengal

Salem Silk - Tamil Nadu

Arani Silk - Tamil Nadu

Molakalmuru Sarees - Karnataka

Ilkat Sarees - Karnataka

Muga Silk - Assam

Orissa Ikat - Odisha

Kancheepuram Silk - Tamil Nadu

Mysore Silk - Karnataka

Chanderi Fabric - Madhya Pradesh

1.11 Policy Initiatives Taken for the Development of Silk Industry

Sericulture is the functional area under the Ministry of Textiles. Some of the recent policy initiatives taken by the Ministry to promote sericulture are as follows.

National Silk Policy 2020

Sericulture is included as agriculture allied activity under RKVY. This enables the sericulturists to avail the benefits of the scheme for the entire sericulture activities up to reeling. The CSB (Amendment) Act, Rules and Regulations have been notified by the Govt. of India to bring quality standards in silkworm seed production.

Forest Conservation Act has been amended to treat non-mulberry sericulture as forest based activity enabling the farmers to undertake Vanya silkworm rearing in the natural host plantation in the forests.

Source: Ministry of Textiles, Government of India

1.12 Research Institutes under Central Silk Board, Govt. of India

Name of the Institution	Located at	Deals with
Central Sericultural Research and Training Institute (CSR&TI)	Mysore, Karnataka	Mulberry sector
	Berhampore, West Bengal	
	Pampore, Jammu and Kashmir	
Central Tasar Research and Training Institute (CTR&TI)	Ranchi, Jharkhand	Tasar sector
Central Muga Eri Research &Training Institute (CMER&TI)	Lahdoigarh, Assam	Eri and Muga sector
Central Silk Technological Research Institute (CSTRI)	Bangalore, Karnataka	Post cocoon sector
Silkworm Seed Technology Laboratory (SSTL)	Bangalore, Karnataka	Seed sector
Central Sericultural Germplasm Resource Centre	Hosur, Tamil Nadu	Germplasm station
Seribiotech Research Laboratory	Bangalore, Karnataka	Seri-biotechnology

References

Silk textile industry in India, silk manufacturers in India - IBEF. India Brand Equity Foundation. (2024, April). https://www.ibef.org/exports/indian-silk-industry.

History of Indian silk. The Indian Silk Export Promotion Council. (n.d.). http://www.theindiansilkexportpromotioncouncil.com/history-of-indian-silk.php.

Vikaspedia domains. Vikaspedia. (2021a, December 15). https://vikaspedia.in/agriculture/farm-based-enterprises/sericulture/sericulture-in-india.

Outcomes Assessment

Part A

Answer the following questions true or false. [1×5=5 Marks]

1. India ranks as world's largest producer of silk. [True/False]
2. In Ancient times China had the monopoly on silk manufacturing. [True/False]
3. Chanderi silk has got its geographic indication from Andhra pradesh. [True/False]
4. Arani silk is from Jammu and Kashmir. [True/False]
5. Central Sericultural Research and Training Institute deals with tasar silk. [True/False]

Part B

Answer the following questions. [2×5=10 Marks]

1. What are the main types of silk worms?
2. Describe briefly about the history of sericulture?
3. Describe about the present status of sericulture in India?
4. What are some of the major constraints faced by sericultural industries in India?
5. What are some of the policy initiatives taken for the development of sericultural industries in India?

Part C

Write a brief note on each of the following. [5×5=25 Marks]

1. Briefly describe the scope and importance of sericulture.
2. Briefly describe the scope and buttons of culture in India and its constraints.
3. Describe the history of sericulture in India and around the world.
4. Briefly describe the five major types of silk worms.
5. Describe above the present status of sericulture in India and around the world.

Part D

Select the correct answer from the four choices given under each question.

[1×10=10 Marks]

1. Which one of the following is the most cultivated species of silkworm in the world.

   a. Mulberry   b. Eri
   c. Tasar   d. Muga

2. Which one of the following countries has the highest production of silk in the world?

   a. India   b. China
   c. Russia   d. Brazil

3. In which of the following countries silk was first discovered?

   a. France   b. India
   c. China   d. United Kingdom

4. Central Sericultural Research and Training Institute deals with which type of silkworm species?

   a. Mulberry   b. Eri
   c. Muga   d. Tasar

5. Silk manufacturing was discovered in China during

   a. 2400 BC.   b. 2000 BC.
   c. 2700 BC.   d. 3000 BC.

6. Muga silkworm is the pride of which state?

   a. Odisha   b. Bihar
   c. Punjab   d. Assam

7. Molakalmuru silk got its G.I. tag in which state?

   a. Karnataka   b. Tamil Nadu
   c. Maharashtra   d. West Bengal

8. Central Tasar Research and Training Institute is located at

   a. Patna, Bihar   b. Bengaluru, Karnataka
   c. Ranchi, Jharkhand   d. None of the above

9. Baluchuri sarees are from which state

   a. West Bengal   b. Bihar
   c. Odisha   d. Jharkhand

10. Yield of silk from India in 2022-2023
    a. 35,000 MT
    b. 47,876 MT
    c. 36,582 MT
    d. 34,567 MT

Answer Key: 1. **a**, 2. **b**, 3. **c**, 4. **a**, 5. **c**, 6. **d**, 7. **a**, 8. **c**, 9. **a**, 10. **c**.

2

Mulberry Varieties

2.1 Introduction

Growing as a perennial tree or shrub, the genus *Morus* (Mulberry) is employed in sericulture and is the only food source for the tamed silkworm ***Bombyx mori* L.** It is also a valuable commercial plant. The mulberry attracts silkworms because it contains elements that cause biting and swallowing. People who live in Karnataka, Andhra Pradesh, the northeastern states and other parts of India rely heavily on sericulture, which grows exclusively mulberries and their variants for the purpose of raising silkworms.

2.2 Mulberry Plant

Mulberry is a plant that belongs to the kingdom Plantae, order Rosales, family Moraceae and genus *Morus*. It grows quickly when young but eventually slows down and rarely reaches a height of 33 to 50 feet. The leaves are alternately arranged, simple often lobed and typically lobed on juvenile shoots rather than on mature trees. The plant can be monoecious or dioecious depending on the species. It produces multiple fruits that are 2-3 cm long and mature fruits that are white or green to pale yellow with pink edges. In most species, the fruits are red when ripened, turning dark purple to black and possessing a sweet flavour.

2.3 Mulberry Species

Moriculture is the term for mulberry plant cultivation. Mulberries come in more than 20 species, four of which are commonly found: *Morus alba, M. indica, M. serrata,* and *M. latifolia* (**Fig. 2.1** to **Fig. 2.4**).

Fig. 2.1: *Morus alba*

Fig. 2.2: *Morus indica*

Fig. 2.3: *Morus serrata*

Fig. 2.4: *Morus latifolia*

The *Morus* was first classified by Linnaeus 1753. Seven species of the genus *Morus* were reported in the book entitled Species Plantarum. They are ***Morus alba, Morus nigra, Morus rubra, Morus tatarica, Morus indica, Morus papyrifera*** and ***Morus tinctoria.***

2.4 Mulberry Varieties

Varieties can be divided into 3 sericultural zones based on the prevailing climatic condition:

2.4.1 Sub-Tropical Sericultural Zone

The region encompasses all of the northern Indian states, which experience harsh weather. This includes the province of Jammu in the state of J & K, which generates the most cocoons overall. From 2 to 3°C in the winter to roughly 24°C in the summer, the temperature varies.

However, the history of sericulture reveals that it was mostly practiced in the temperate zone, which will be covered after you have a better understanding of the sub-tropical zone. For a very long period, native or wild mulberry cultivars have allowed sericulture to thrive in the subtropical zone. A typical local mulberry, tiny and thin-leafed and poor yielder. Later on, better cultivars were discovered and made available for commercial use, including **Sujanpur local** and other local types. Numerous kinds from around the nation have been released since the 1970s. The cultivars **S-146, Tr-10, and Chinese White** are deemed better appropriate for the sub-tropical sericulture zone and have yielded satisfactory results for the local sericulturists. In sub-tropical areas, the following qualities are preferred in mulberry varieties:

i. Can be trained as a tree

ii. Early sprouting in nature

iii. Late leaf fall

iv. Less fruiting

v. Drought tolerance

vi. Resistant to diseases and pests

2.4.2 Temperate Sericultural Zone

During the winter, snowfall falls in these places. The Kashmir Valley and the upper regions of Jammu, Himachal Pradesh and Uttaranchal make up the majority of this zone. However, the Kashmir valley is the exclusive location of the majority of the zone's sericultural activity. Additionally, sericulture has flourished in this area thanks primarily to indigenous cultivars like **Chattatul, Botatul, Zagtul,** etc.

The 1950s saw the beginning of the process of improving the introduction of mulberry varieties in this region. The behaviour of mulberry types varies according to the type of climate. It is worth noting that Japan accounts for around 60% of the zone's overall mulberry wealth. Popular Japanese types in this area include **Ichinose, Goshoerami, KNG, Kakuso-27** and others. Another significant factor in the mulberry wealth in this zone is wild mulberry.

For mulberry variants in temperate climates, the following desirable traits have been identified: Quick growth, Early sprouting, Steep branching, Frost resistance, Farmer acceptance, Reduced fruiting, Adaptability for tree training.

2.4.3 Tropical Sericultural Zone

Odisha, Maharashtra, Karnataka, Andhra Pradesh, Tamil Nadu, Telangana, Kerala, parts of North Eastern India and West Bengal make up the majority of this zone. Karnataka is an exclusive location for sericultural activities and research for this zone. There Indigenous/Traditional varieties have been flourished in this zone like ***M. indica,*** **Berhampore local, Mysore local, Kanva-2** etc. These varieties can be used both in rainfed and irrigated conditions giving an optimum yield. Some of the improved varieties are **Vishala, Vishwa, Suvarna-2 and Suvarna-3** etc. These varieties give higher yield.

Some other varieties are:

A. Improved mulberry variety: **'Viswa'** (Dehra Dun, DD), **'TG' (Thlaghattapura), Suvarna-1, Suvarna-2, Suvarna-3.**

B. High yielding mulberry variety: '**Vishala**'.

C. Mulberry genotypes for cultivation as small trees under rainfed conditions: **S146, S13, Vishwa.**

D. Potential mulberry genotype: **S1635** for irrigated conditions.

2.5 Uses of Mulberry

The ripe fruit is used in pies, tarts, wines and teas. It is edible. The strongest-tasting mulberries are those of the red mulberry, which is native to eastern North America, and the black mulberry, which is native to south west Asia. The leaves and fruits are offered for sale as dietary supplements in a variety of forms. The plant's green sections and unripe fruit have an enticing white sap. The leaves of mulberries, especially white mulberries, are crucial to the ecology because they are the only food supply for silkworms, which produce silk from their pupae or cocoons. Pigments called anthocyanins have the potential to be used as natural food colouring agents as well as dietary modulators of disease-causing processes. Food colourants are becoming more and more important in the food sector as a result of rising demand. The vibrant hues of fresh food plants, including orange, purple, black and blue, are produced by anthocyanins. They are easily extracted and soluble in water. India has long grown mulberries, the only plant used as nourishment for silkworms. Mulberries are found all across the world, with various species being found in large numbers all over the nation. The leaves offer more promise for treating a variety of illnesses in addition to providing food for silkworms.

References

Mulberry Breeding and Genetics - Karnataka State Sericulture Research and Development Institute.

https://kssrdi.karnataka.gov.in/page/Contributions/Moriculture%20Technologies/MulberryBreeding and Genetics/en.

Mulberry production. Central Sericultural Research & Training Institute (CSRTI), Mysore. (n.d.). http://www.csrtimys.res.in/structure/mulberry-production.

Outcomes Assessment

Part A

Answer the following questions true or false. [1×5=5 Marks]

1. Mulberry belongs to family Rutaceae. [True/False]
2. The cultivation of mulberry plants is known at sericulture. [True/False]
3. *Morus* was first classified by Aristotle. [True/False]
4. Chattatul is a tropical zone variety of mulberry. [True/False]
5. Chinese white is a variety of subtropical zone. [True/False]

Part B

Answer the following questions. [2×5=10 Marks]

1. Species of mulberry and their characteristics.
2. Varieties of Tropical Sericultural Zone.
3. Varieties of Sub-Tropical Sericultural Zone.
4. Varieties of Temperate Sericultural Zone.
5. Uses of mulberry plant.

Part C

Write a brief note on each of the following. [5×4=20 Marks]

1. Mention some of the varieties of Tropical Sericultural Zone and their characteristics.
2. Mention some of the varieties of Sub-Tropical Sericultural Zone and their characteristics.
3. Mention some of the varieties of Temperate Sericultural Zone and their characteristics.
4. Write about the characteristics of different mulberry species.

Part D

Select the correct answer from the four choices given under each question.

[1×5=5 Marks]

1. Mulberry belongs to the family
    a. Moraceae
    b. Asteraceae
    c. Euphorbiaceae
    d. Rutaceae
2. What type of colouring pigments are present in mulberry?
    a. Cryptoxanthin
    b. Anthocyanin
    c. Carotene
    d. None of the above
3. Which of the following is an indigenous variety of mulberry?
    a. Vishwa
    b. Kanva-2
    c. Vishala
    d. Suvarna-2
4. Which of the following cultivar is an indigenous variety grown in subtropical sericultural zone of India?
    a. Sujanpur local
    b. Kanva-2
    c. Vishala
    d. Botatul
5. Potential mulberry genotype for irrigated conditions?
    a. S1635
    b. Vishala
    c. Kanva -2
    d. S13

Answer Key: 1. **a**, 2. **b**, 3. **b**, 4. **a**, 5. **a**.

3

Host Plants for Non-Mulberry Silkworms

3.1 Introduction

Non-mulberry plants, sometimes referred to as **Vanya plants**, are important to the sericulture sector because they give silkworms—which are normally fed mulberry leaves—an alternate source of nutrition. Many species, including castor (*Ricinus communis*), oak (*Quercus* spp.) and arjun (*Terminalia arjuna*), are among these non-mulberry plants. These plants are a food source for a distinct species of silkworm known as Eri, Tasar and Muga silk, which are distinct from the common mulberry silkworm, *Bombyx mori*. These silks are prized for their unique properties, hues and textures which add to the variety and depth of silk goods on the market.

3.2 Host Plants

3.2.1 Host Plants for Tasar Silkworms

The traditional tasar and the oak tasar are now two separate entities within the Indian tasar culture. ***Terminalia*** and ***Shorea*** dominate the traditional tasar tract, with ***Quercus*** oaks (as the name suggests) dominating the latter. Although tasar silkworms are actually polyphagous, there is some specificity in terms of their affinity for a given feeding source.

Taxonomy: *A. mylitta* feeds primarily on Asan (*Terminalia tomentosa*, **Fig.3.1**), Arjun (*Terminalia arjuna*, **Fig.3.2**) and Sal (*Shorea robusta*, **Fig.3.3**), while *A. proylei* is mainly reared on Uyung (*Quercus serrata*, **Fig. 3.4**), Banj (*Quercus incana*, **Fig. 3.5**), Sahi (*Quercus dealbata*, **Fig. 3.6**) and Moru (*Quercus himalayana*, **Fig. 3.7**). Being important for tasar industry, these three genera (*Terminalia, Shorea* and *Quercus*) the silkworm feeds on Phutuka (*Melastoma*, **Fig. 3.8**) and Bogori (*Zizyphus jujuba*, **Fig. 3.9**).

Fig. 3.1: Asan (*Terminalia tomentosa*)

Fig. 3.2: Arjun (*Terminalia arjuna*)

Fig. 3.3: Sal (*Shorea robusta*)

Fig. 3.4: Uyung (*Quercus serrata*)

Fig. 3.5: Banj (*Quercus incana*)

Fig. 3.6: Sahi (*Quercus dealbata*)

Fig. 3.7: Moru (*Quercus himalayana*)

Fig. 3.8: Phutuka (*Melastoma*)

Fig. 3.9: Bogori (*Zizyphus jujuba*)

Fig. 3.10: Castor (*Ricinus communis*)

Fig. 3.11: Soalu (*Litsaea polyantha*)

Fig.3.12: Som (*Machilus bombycina*)

3.2.1.1 Terminalia spp. (Order: Myrtales; Family: Combretaceae)

Distribution: The lower hills and valleys of northern India are home *to T. tomentosa. T. arjuna is* more prevalent in southern Bihar and Madhya Pradesh in small streams.

Plantation: Tasar larvae are traditionally fed on erratically dispersed food plants in paddy field bunds or woodlands posing management risks that lead to significant losses from illnesses and pests.

Soil: Reddish or well-drained lateritic soils are ideal for *Terminalia tomentosa* and related species. It grows well in soils with loam, sandy loam and even clayey loam.

Land Preparation: The ground is thoroughly prepared by disking, levelling, ploughing and terracing as needed. Before transplanting, 2×2×2-foot pits are dug. In four-year-old plantations with four or five spacings, the yield of leaves per hectare was 136.52, 97.25 and 62.44 quintals, respectively. Thus, if planted for tasar culture, plantations with tighter spacings can result in a higher income for the rearer.

Transplanting: A mixture of earth, FYM/compost, 100–200 g of karanj cake, bone meal, wood ash and 5% Aldrin dust are placed into the transplantation-ready pits. Strong, robust seedlings are transplanted in June or July and then given one rose can watering.

Manure and Fertilizer: Almost minimal manuring is done on tasar food plants. However, some recent studies in this area have shown a notable improvement in leaf output. The recommended dose per plant is 20 kg of FYM plus 14 g of N and 7 g of P_2O_5 chemical fertiliser. Manures are applied as basal dose.

Training and Pruning: The plants that provide food for tasars often grow to a height of 4-6 feet. However, plants with well-spread crowns and dense foliage that are between 10 and 12 feet tall are perfect for rearing. Thus, it is imperative to start training the plants at a very young age. Three training strategies are typically applied to horticultural trees: the Central Leader, Open Centre and Modified Leader systems. The central leader system preserves the dominance of the main trunk while preventing the lateral branches from spreading and leaving one or two branches at constant heights, giving the plant a linear form with short dense branches.

3.2.2 Host Plants for Eri Silkworms

Primarily, the Eri silkworm consumes *Ricinus communis,* also known as the castor oil or bharenda plant (**Fig.3.10**).

Distribution: *R. communis* is grown in various countries, such as Brazil, the United States of America, Mexico, Sudan, Egypt, India, Burma and Sri Lanka. It is widely dispersed throughout India and is mostly utilized for eri cultivation in the states of Assam, Tripura, West Bengal, Bihar, Meghalaya, Arunachal Pradesh and Manipur where it is grown on hillsides, bunds and other wastes as a self-generating crop.

The following is a description of *R. communis* (Order: Euphorbiales; Family: Euphorbiaceae) for cultivation:

Sowing of Seeds: Castor can be cultivated as an annual or as a plant with leaves that are either violet or pale green. Both work for feeding silkworms that are eri. Grows well in September. The factors of heat, light, moisture and aeration have a major impact on the germination of seeds and the growth of castor plants.

It must therefore be grown in the same manner as other garden crops. Occasionally, nine to ten kilograms of seeds are sown per acre. However, the best way is to plant two to three seeds in each hole, spaced either 3' to 4' apart or 5' to 6' apart in rows. When given the ideal circumstances of warmth, light, moisture and airflow seeds typically sprout within 7 to 10 days.

Soil and Manure: Trees spaced 12 feet by 12 feet apart can produce up to 20,000 kg of leaves per acre. Light clay soil is necessary for castor tree cultivation to be successful. But clay soil is never better than light sandy soil. Even though the soil needs potash, it can be restored by adding ash from burned weeds, leaves, stalks and stems.

You can use castor and mustard oil cakes as manure. Every two or three years, when manure containing oil is used, a lime dose is required. In certain areas, well-decomposed farm yard manure is the best and most affordable. You can apply inorganic manures as a top dressing at the start of the crop or in the middle of it.

Methods of Harvesting: Due to uneven capsule maturation across the fruiting branch or spike the castor seeds never fully ripen at one time. Thus, it is time to harvest the crops when a few dried fruits or capsules are visible. It will not be beneficial to wait to harvest the castor seed on this account. If the harvest is postponed until the fruits on the vines dry, mature capsules will release their seeds. The cost of collecting will be somewhat high and more work will be needed for the frequent pickings if castor seeds are picked by hand while they are dry. It follows that growing strains whose seed maturity occurs at a time is an apparent solution to this issue.

3.2.3 Host Plants for Muga Silkworm

The polyphagous Muga silkworm insect consumes the leaves of several plants. The two most significant are Soalu (*Litsaea polyantha,* **Fig. 3.11**) and Som (*Machilus bombycina,* **Fig. 3.12**), which are widely distributed throughout North Eastern India, especially in the Brahmaputra river, the foothills of Naga, Khasi and Garo hills.

Nearly all muga silkworm food plants thrive in warm, humid climates with plenty of rainfall. The majority of muga food plants are members of the Lauraceae family (order: Laurales). According to the survey, the majority of plant species are found in their natural habitat in the sub-Himlayan hill ranges, especially in the North Eastern region.

Topography and Soil: Sandy loam soil is ideal for muga plant growth. The pH of the soil in which Muga food plants are grown ranges from 4.5 to 5.0, making it acidic. Muga plantations can be established on high, sloping areas of land with adequate drainage.

3.2.3.1 Nursery Techniques and Propagation Methods for Growing Muga Food Plant

Propagation via Seeds: In the wild, seeds are used to grow som and soalu plants. While soalu seeds are collected in May and June, som seeds are gathered in March and April. After kneading the seed two or three times in water to remove the pulp, the seed is dried for a few hours in the shade. The seeds have a relatively limited shelf life.

In the prepared beds, seeds are spread in rows at a depth of approximately 2 cm, with a spacing of 15 cm inside each row and 15 cm between rows. Four weeks after sowing, germination begins. In a similar manner, polythene tubes can be used to nurture seedlings. The seedlings are kept for around three months, or until they reach one foot, in nursery or polythene tubes. The nursery's seedlings are moved into 1-by-l-by-l-foot pits and fertilized with 0.5 cubic feet of farm yard manure and 1 g aldrin (5%) dust. Filling the pit with dirt, manure and insecticides should be done properly before planting the seedling. Therefore, in a square system of plantation, a spacing of 4 × 4 metres for som and 5 × 5 metres for soalu is suitable for accommodating 625 and 400 plants/ha respectively.

Site and Layout Selection: Muga food crops should not be located in low-lying areas or have areas where water stagnates. High and well-drained ground is ideal. In a forest area, the ground must be cleared of trees, shrubs and other plants. Next, the area needs to be divided into manageable plots and all of the tree roots need to be entirely removed.

Choosing Edible Plants: Despite being polyphagous muga silkworms choose som and soalu as their primary host plants. Because som grows evergreen and has a consistently higher effective rate of rearing, it is the perfect plant for muga culture. Cocoons generated by som-fed worms have better quantitative characteristics than those produced by soalu-fed worms, such as shell weight, silk ratio and reliability. On the other hand, muga silkworms that are fed soalu leaves yield healthy moths.

Digging and Filling of Pits: The following step is pit digging and pit filling, which comes after site selection, plant selection and layout. One cubic foot (1 × 1 × 1) pits are excavated in the dry spring months of February and March. To completely dry up the weeds and their seeds, the excavated earth can be left out in the sun. The dirt that was excavated from the hole needs to be well combined with 10 grams of Aldrin (5%) dust and farm yard manure (FYM). The mixture is placed into the ready-made trenches and stored until the monsoon season (July to September) arrives.

Planting Season: The best time to plant muga food plants is during the monsoon season, which runs from July to September and experiences considerable rainfall. During this time, the plants stay in an active growth stage. This is the season when the roots properly develop.

Manuring: Farm yard manure is applied @ 0.5 cft/plant, and starting in the second year, FYM is applied @ 1 cft/plant. This needs to be done for a minimum of five years. Frequently, a circular trench 6 to 9 inches deep is excavated around the plant, especially while the seedlings are small.

Training and Pruning: Pruning and training of muga host plants properly is necessary. The goal is to: (I) grow foliage; (II) eliminate sick or dead sections; (III) give the plant a suitable form; (IV) regulate the plant's height; and (V) create a functional balance between the various organs of the plant.

Training of Seedlings: The apical tips of the seedlings with three to four leaves are manually plucked when they reach a height of twelve inches. When the axillary buds open, new branches will grow in roughly 20 days. Using this method, seedlings with four to five branches may be produced.

Training of Old Tree: Som and Soalu plants can reach heights of 50 to 60 feet in the wild. As a result, raising issues are inevitable due to such towering plants. Trees that are 15 to 20 feet tall, have a spreading crown, and produce lush foliage are perfect for handling and raising muga silkworms.

Maintaining four to five branches at the base of an older tree and allowing two to three branches to sprout on each of these branches is the optimal approach for training older trees. To ensure that an umbrella-shaped crown is firmly in place, this procedure should be followed year after year. This raises the potential for raising muga plants which results in more yield in leaves.

References

Non mulberry food plant cultivation. - H.B. Mahesha. (n.d.-c). https://hbmahesh.weebly.com/uploads/3/4/2/2/3422804/non_mulberry_food_plant_cultivation.

Outcomes Assessment

Part A

Answer the following questions true or false. [1×4=4 Marks]

1. Mulberry plant is commonly used as a host in Muga silkworm. [True/False]
2. Castor plant is commonly used as a host in Tasar silkworm. [True/False]
3. Tasar silkworms commonly feed on *Terminalia* spp. [True/False]
4. *Terminalia tomentosa* grows well in black cotton soil. [True/False]

Part B

Answer the following questions. [2×4=8 Marks]

1. What are the common host plants for Eri silkworm and their characteristics?
2. What are the common host plants for Muga silkworm and their characteristics?
3. What are the common host plants for Tasar silkworm and their characteristics?
4. Write down the cultivation practices of *Terminalia* spp.?

Part C

Write a brief note on each of the following. [5×3=15 Marks]

1. What are the common host plants for Eri silkworm with their characteristics and cultivation practices?
2. What are the common host plants for Muga silkworm with their characteristics and cultivation practices?

3. What are the common host plants for Tasar silkworm with their characteristics and cultivation practices?

Part D

Select the correct answer from the four choices given under each question. [1×4=4 Marks]

1. Which of the following is the host plant for Muga silkworm?
   a. Terminalia and shorea b. Som and soalu
   c. Castor d. Mulberry
2. Which of the following is the host plant for Eri silkworm?
   a. *Terminalia and shorea* b. Som and Soalu
   c. Castor d. Mulberry
3. Which of the following is the host plant for Tasar silkworm?
   a. Terminalia and Shorea b. Som and soalu
   c. Castor d. Mulberry
4. Best planting season for som and soalu?
   a. Apr-Jun b. Dec-Feb
   c. Jul-Sep d. Oct-Dec

Answer Key: 1. **b**, 2. **c**, 3. **a**, 4. **c**.

4

Preparation of Land, Planting Material and Planting of Mulberry

4.1 Introduction

The yield of the mulberry crop is a significant component in determining productivity and consequently, profitability in sericulture. The two most crucial goals—increasing cocoon output per hectare and lowering production costs—will be achieved through optimizing mulberry leaf yield per unit area. Therefore, ensuring that he receives the highest leaf yield possible from his mulberry crop ought to be the main goal of every sericulturist. Furthermore, it should be understood that every action made to increase leaf production also serves to enhance leaf quality, so securing insurance against losses to the cocoon crop later on in the silkworm raising process.

4.2 Soil and Climate

Except for extremely steep terrain, mulberry may be grown practically on any type of terrain. However, when it is grown on either level ground or gently sloping or undulating terrain, good growths are obtained. More sloping or steep terrain calls for careful consideration of appropriate soil conservation techniques including bench terracing, contour planting and contour drainage. Although mulberry thrives in a variety of soil types, loamy to clayey loam soils are ideal for growth. Mulberry plants can withstand somewhat acidic soil conditions. However, if the pH of the soil is too low, lime or dolomite should be applied as a necessary corrective measure. When it comes to alkaline soil Gypsum application in the soil is the best option for adjusting the soil's alkalinity. Given that mulberries have deep roots, the soil should be sufficiently deep—up to two feet, at the very least. Mulberries grow well up to an elevation of around 4,000 feet after which their growth will be slowed by the lower temperatures.

4.3 Land Preparation

Utilizing the pre-monsoon showers in April and May, the land should be thoroughly ploughed with a heavy mould board plough to a depth of 12"–

15" (30–35 cm) to loosen the soil before planting. After that, the ground may be lightly ploughed once or twice to achieve fine-tilled soil. Subsequently, a minimum of 10 tonnes per hectare of organic manure, such as compost or animal manure, should be applied for rainfed mulberry trees and 20 tonnes per hectare for irrigated trees as a basal dose. Lastly, the ground should be levelled and prepared for planting during the June–July monsoon rains by ploughing the manure into the soil properly.

In general, mulberry plants that are rainfed should use a pit planting system with wider spacing, whereas mulberry plants that are irrigated should use a row planting system with closer spacing. Pits should therefore be excavated at a spacing of 3' × 3' (0.9 m × 0.9 m) in order to sow mulberries under rainfed conditions. The dimensions of the pits should be 3.11' × 1.25' (95 cm × 35 cm) with a minimum depth of 1.25'×1.25' (35 cm). Cuttings or rooted saplings are placed in these pits which are filled with dirt and ideally some cattle manure.

4.4 Planting Material and Planting

Mulberries may readily root in tropical climates, such as those found in South Africa, and can thus be propagated through cuttings with little effort and expense. The cuttings must be made from brown, hard wood branches that are at least (10-12 mm) in diameter and 4–8 months old. The cuttings ought to have three buds at the very least and measure at least 7-8" (18-20 cm). With a sharp knife, the cutting ends should be neatly sliced, free of splits and peeling bark.

A. Thin and short cuttings are not appropriate for planting.

B. When improper cuttings are planted, there will be little to no root production and growth (although they may sprout).

C. Proper cutting technique: 1/2" diameter and 7-8" length.

D. Vigorous root production and growth upon planting the appropriate variety of cutting. Errors are frequently committed in the planting material selection process, leading to poor plant establishment with numerous failures and resulting gaps. Avoid cuttings that are either tiny in diameter or green in colour since their chances of success are low. When rooting the cuttings extreme caution must be taken.

Direct planting of cuttings is extremely dangerous in areas without access to compost or animal dung. In such circumstances, it will be essential to cultivate rooted plants in nurseries and move 2.5–3 month-old rooted plants, with a stem thickness of roughly 10 mm and a growth of approximately 3 feet onto the main field. When moving nursery-raised plants, it is crucial to make sure the original cutting from which the plants grew is buried deeply in the ground—at least one to two inches below the surface—and that the surrounding soil is compacted. Better plant anchoring is ensured by doing this, using either cuttings or plants grown in nurseries for all the fresh plantings.

4.5 Spacing

The goal of raising mulberry plants with a stronger frame in rainfed mulberry gardens is to make them more resilient to the current drought conditions. As a result, as is now the case, the space should be at least 3' × 3' (0.9 m × 0.9 m). Cuttings should be planted in "threes" in the trenches designated for that purpose spaced six inches (15 cm) apart to make an equilateral triangle. Plants that have been rooted in nurseries may be transferred as single plants. When growing mulberry in an irrigated area leave 2 feet (0.6 m) between rows and 9-10" (23–25 cm) within each row.

4.6 Manure and Fertilizer

Farmyard manure can be applied once every six months during bottom or middle pruning, at a rate of four tones per acre. Use compost, oil cakes, vermicompost, green leaf manure, poultry manure and silt from the reservoir in place of farm yard manure if it is unavailable. When fertilizers are used at the proper dosage very good mulberry yields can be achieved. Use 300:120:120 kg of nitrogen, phosphorus and potash fertilizers in five separate doses per year when using the shoot harvest method. Apply 350:140:140 kg of nitrogen, phosphorous and potash fertilizers in five separate dosages per year according to the Jaika suggested system. Use 280:120:120 kg of nitrogen, phosphorus and potash fertilizers in six separate doses per year when using the leaf harvest method. In regions with hills use 250:100:100 kg of NPK, Phosphorus: Potash fertilizers in five divided dosages a year and in rain-fed mulberry gardens containing 100:50:50 kg of NPK. It is advised to apply potash, phosphate and nitrogen fertilizers in two separate dosages each year.

4.7 Irrigation

Irrigate once in a week @ 80-120 mm. Wherever there is a shortage for water farmers can go for drip irrigation by which 40% irrigation water can be saved.

References

Anonymous. 1975. Textbook of Tropical Sericulture. Japan Overseas Cooperation Volunteers, 4 -2-24 Hiroo, Sibuya-Ku, Tokyo, Japan. pp. 1-194.

Dandin, S.B., Jayaswal, J. and Giridhar, K. 2003. Handbook of Sericulture Technologies (3rd Edn.). Central Silk Board (Ministry of Textiles – Govt. of India, BTM Layout, Madiwala, Bangalore. pp. 1-277.

Dandin, S.B., Rajan, R.K., Mishra, R.K., Singh, G.B. and Das, P.K. 2002. A Model for Commercial Chawki Rearing (Technical Booklet – 3). Central Sericultural Research & Training Institute, CSB (Ministry of Textiles – Govt. of India, Manandavadi Road, Mysore. pp. 1-23

Kawakami, K. and Yanagawa. 2003. Illustrated Working Process of New Mulberry Cultivation Technology (English – Kannada). JICA, India Project for Strengthening Extension System for Bivoltine Sericulture in India. pp. 1-78.

Outcomes Assessment

Part A

Answer the following questions true or false. [1×4=4 Marks]

1. Mulberry plant grows well in a low elevation area. [True/False]
2. Mulberry plant is commonly propagated through seeds. [True/False]
3. Mulberry plants grows well in a loamy to clay loamy soils. [True/False]
4. Requirement of nitrogen in mulberry cultivation is less in comparison to potassium and phosphorus. [True/False]

Part B

Answer the following questions. [2×4=8 Marks]

1. What are the soil and climate requirements for mulberry cultivation?
2. Process of land preparation for mulberry cultivation.
3. What is the fertilizer requirement for mulberry cultivation?
4. Describe about the planting material and planting of mulberry.

Part C

Write a brief note on each of the following. [5×2= 10 Marks]

1. Briefly describe about the soil, climatic condition, land preparation, planting and planting material for mulberry cultivation.
2. Describe about the manure and fertilizer requirement, plant spacing and irrigation requirement for mulberry cultivation.

Part D

Select the correct answer from the four choices given under each question. [1×4=4 Marks]

1. What is the age of the cutting which will be used as planting material?

   a. 3 months b. 9-12 months

   c. 4-8 months d. 1.5 years

2. Mulberry grows well around an elevation of

   a. 3000m b. 1000m

   c. 2000m d. 4000m

3. How many cuttings should be planted in one trench?

   a. 1 b. 2

   c. 3 d. 4

4. By the use of drip irrigation what percentage of irrigation water can be saved?

   a. 40 b. 50

   c. 80 d. 90

Answer Key: 1. **c**, 2. **d**, 3. **c**, 4. **a**.

5

Pruning, Harvesting and Storage of Mulberry

5.1 Pruning

One of the most crucial aspects of mulberry farming is pruning. It involves methodically cutting off some branches in order to give the mulberry plants a manageable size and form as well as to encourage the bud to sprout and develop new branches. The primary goal of pruning is to increase leaf output and quality. In addition, it is used to harvest the shoots for silkworm rearing and to do various inter-cultivation tasks including weeding, ploughing, applying manure and fertilizer and proper timing of the rearing of silkworms. To prune without breaking the bark or stem, use a sharp secateur, pruning saw or sickle. The plants in rainfed environments are only ever trimmed once a year, at a height of 10–20 cm above the ground level in June or July when the monsoon season begins. After that the plants are let to grow into bushes with many branches till the next year. Leaf harvesting is limited to the leaf picking method.

a) **Pruning Types:** Pruning is categorized according to the height of the stem at which it is cut. Bottom pruning is the practice of cutting the main stem at ground level after each harvest so as not to allow a stump to emerge. When low-cut pruning is used, the main stump of the plant is developed by cutting it 10 to 30 centimetres above the ground. At around 60 and 100 centimetres above the ground, medium and high cut pruning is performed respectively.

   Primarily, there are two kinds of pruning:

   i) **Fist Form:** This form has branches that are clipped at the same height at the base. Since no new branches are permitted to grow into new stumps, the plant widens and takes on the form of a fist.

   ii) **Non-Fist Form:** Two to three buds remain at the base of the shoot after the branches are chopped. Because of this after each shoot harvest the plants grow taller. A non-fist method of pruning encourages the growth of extra shoots.

b) **Thinning of Branches**: Removing extra branches from a plant to encourage healthy growth and leaf production is known as branch thinning. Pruning mulberry plants causes a large number of buds to sprout. However, after a few days only a small percentage of these buds mature into strong healthy shoots leaving the majority of the sprung buds stunted and feeble. Raising silkworms is not a good use for such feeble shoots. Furthermore, the competition between numerous developing shoots leads to an incorrect distribution of plant nutrients among all the shoots which ultimately results in the development of weak shoots. Therefore, it is crucial to limit the number of branches (10–12) per plant by trimming off any extra branches after 30–35 days.

5.2 Harvesting in Mulberry

Since gathered leaves are required to raise silkworms, harvesting leaves is a crucial part of mulberry agriculture. Harvesting leaves can be done in two major ways: (i) picking leaves by hand and (ii) harvesting entire shoots.

a) Leaf Plucking: Using this technique, leaves are collected by plucking each one separately. The benefit of this procedure is that the leaf is harvested based on the silkworm's stage of growth. Mature leaves are used to nurture late-stage silkworms, while tender leaves are collected to feed the younger silkworms. This method's drawback is that it requires a lot of manpower which drives up the cost. After the leaves on the main branches are harvested, the terminal buds are cut off and the auxiliary buds are allowed to grow into numerous lateral branches in southern states, particularly when they receive rain. Picking is then done to gather the leaves from the lateral branches. Six harvests, spaced 55–60 days apart can be made from a garden using the leaf picking method in a year. Typically, following the yearly bottom trimming, the first leaf harvesting occurs 8–10 weeks later with subsequent harvests occurring 6–8 weeks apart.

b) Shoot Harvesting: This technique involves cutting the entire shoot at a height of 20 to 25 cm above the ground, using the entire shoot for raising silkworms. With this system, a crop cycle of 70 days is followed by five harvests in a year one for each crop. With this procedure, plant pruning and leaf harvesting are done at the same time. It is best to use a pruning saw or secateur for pruning, and be sure not to harm the main stump's bark.

5.3 Transportation and Storage

Shortly after being harvested and transported to the rearing house, the leaf or shoot begins to lose moisture. This is especially true in situations where the humidity is low and the air temperature is high. The leaf is therefore unsuitable for raising silkworms. To prevent moisture loss, it is crucial to handle the leaf carefully while it is transported from the field to the rearing house. When leaves are picked for the purpose of raising young silkworms this becomes even more crucial. It is preferable to build the rearing house or shed close to the mulberry field for convenient and speedy transportation of leaves and shoots while maintaining the moisture content of the leaves. This lowers transportation costs and guarantees a supply of fresh leaves for silkworm rearing. Harvesting of leaves and shoots is best done in the cooler parts of the day and they should be carried to the rearing house in baskets covered in moist gunny cloth. To prevent moisture loss during transportation of large quantities of leaves or shoots, cover them with damp gunny cloth or polythene sheets. Since the silkworms are fed two or three times a day, it is critical to keep the leaf or shoot in a clean, damp and cool environment. While the shoots are preserved by covering with damp gunny cloth it is preferable to preserve the leaves in the leaf chambers.

References

Anonymous. 1975. Textbook of Tropical Sericulture. Japan Overseas CooperationVolunteers, 4 -2-24 Hiroo, Sibuya-Ku, Tokyo, Japan. pp. 1-194.

Dandin, S.B., Jayaswal, J. and Giridhar, K. 2003. Handbook of Sericulture Technologies (3rd Edn.). Central Silk Board (Ministry of Textiles – Govt. of India, BTM Layout, Madiwala, Bangalore. pp. 1-277.

Dandin, S.B., Rajan, R.K., Mishra, R.K., Singh, G.B. and Das, P.K. 2002. A Model for Commercial Chawki Rearing (Technical Booklet–3). Central Sericultural Research & Training Institute, CSB (Ministry of Textiles–Govt. of India, Manandavadi Road, Mysore. pp. 1-23

Kawakami, K. and Yanagawa. 2003. Illustrated Working Process of New Mulberry Cultivation Technology (English–Kannada). JICA, India Project for Strengthening Extension System for Bivoltine Sericulture in India. pp. 1-78.

Outcomes Assessment

Part A

Answer the following questions true or false. [1×4=4 Marks]

1. The rainfed trimming in mulberry plant is trimmed during October to January. [True/False]
2. Mulberry plant is commonly propagated through seeds. [True/False]

3. Shoot harvesting requires more manpower than leaf plucking. [True/False]
4. In fist form of pruning the plant is pruned at the same height from the base. [True/False]

Part B

Answer the following questions. [2×3=6 Marks]

1. What are the types of pruning in mulberry cultivation?
2. What are the types of harvesting in mulberry cultivation?
3. Describe about the transport and storage of mulberry leaves.

Part C

Write a brief note on each of the following. [5×2= 10 Marks]

1. Briefly describe the pruning and thinning practices for mulberry cultivation.
2. Describe the methods of harvesting, transport and storage of mulberry leaves.

Part D

Select the correct answer from the four choices given under each question.

[1×4=4 Marks]

1. What is the age of the cutting which will be used as planting material?
   a. 3 months   b. 9-12 months
   c. 4-8 months   d. 1.5 years
2. In which form of pruning two to three buds remain at the base of the shoot after the branches are chopped?
   a. Fist form   b. Non-fist form
   c. Both a. And b.   d. None of the above
3. Shoot harvesting involves the cutting of entire shoot from a height of?
   a. 30 cm   b. 20-25 cm
   c. 35 cm   d. 14 cm
4. Medium and high pruning are performed at a height of?
   a. 40 cm   b. 50 cm
   c. 60-100 cm   d. 20-40 cm

Answer Key: 1. **c**, 2. **b**, 3. **b**, 4. **c**.

6

Mulberry Pests and Diseases

6.1 Introduction

India's mulberry plants, which are vital to sericulture, are susceptible to a range of pests and diseases that can harm them and reduce their productivity. The Leaf Roller, Mealy Bug, Whitefly, and Stem Borer are important pests that harm leaves and stems and lower plant yield. Plant degradation and leaf discolouration are caused by common diseases such as Powdery Mildew, Leaf Spot, Root Rot, and Bacterial Blight. To preserve healthy mulberry harvests and continue to produce silk, effective management requires the application of Integrated Pest Management (IPM) techniques, which include biological, chemical, and cultural approaches.

Mulberry is also prone to a number of illnesses brought on by bacteria, mycoplasma toxins, and fungi. It is simple for a perennial agricultural plant to become vulnerable to different diseases. Similar to this, pest insects have a plenty of food available to them all year round to aid in reproduction. Therefore, developing high-yielding types through plant breeding programs is just as vital as preserving mulberries from disease and insect attacks. The diseases listed below are accompanied with control methods. Adopting appropriate management techniques will help prevent pests and diseases while producing high-quality and productive mulberry leaf.

6.2 Mulberry Pests

1. Pink Mealybug

Adults and nymphs suck the cell sap from fragile buds and leaves. There is a significant decrease in plant height, leaf production, and nutritional value (**Fig. 6.1**).

Fig. 6.1: Pink mealybug

Signs and Symptoms

Apical sprout malformation, slowed growth, leaf curving and wrinkling, and darkening of the leaf's colour are all present. When an infestation is strong, the leaves turn pale yellow. Affected areas start to crack. The collective term for the symptoms is Tukra (Bushy top) illness.

Management

After chopping off the afflicted shoots, burn the area with either Dichlorvos 76 WSC @ 2 ml/l or Fish Oil Rosin Soap (FORS) @ 40 g/l (safe waiting period: 15 days). Releasing 1000 beetles/ha of ***Scymnus coccivora*** and 750 beetles/ha of ***Cryptolaemus montrouzieri***

2. Papaya Mealy Bug

Margatus paracoccus Granara de Willink and Williams (Pseudococcidae: Hemiptera)

It appears to have been brought to India and is a new record with an unusual origin. It is an invasive pest that affects a wide range of industrial crops (**Fig. 6.2**).

Fig. 6.2: Papaya mealy bug

Signs and Harm

First, it affects the apical portions of the plant, and then it spreads throughout the entire plant, even harming the woody areas.

Affected portion's deformity as a result of the poison administered during feeding slowed plant growth and leaf yellowing, Plants and leaves with sooty mould as a result of the pest's honeydew secretions, Ant activity nearby that contributes to the mealy bug's spread removing the plant entirely in the event of a severe infestation.

Parasitoids

With the assistance of USDA-Animal and Plant Health Information Services, NBAII has successfully imported three Encyrtid parasitoids from Puerto Rico: ***Acerophagus papaya*, *Pseudleptomastix mexicana*,** and ***Anagyrus loecki*.** The three parasitoids have also successfully completed all required safety and specificity tests in the quarantine facility. In the lab, the parasitoids can be effectively replicated on *Paracoccus marginatus* colonies cultured on potato shoots and sprouts.

Method for traditional biological regulation - In hotspot areas infected with pests, 500 parasitoids of each foreign species are released as an inoculant; releases may be repeated if needed. Avoiding the use of chemical pesticides can help conserve the released parasitoids and naturally occurring predators like spalgis and coccinelids.

3. Leaf Webber

Fig. 6.3: Leaf webber

Type of damage

Larvae defoliate the mulberry plants. Reduction in leaf yield.

Symptoms

Folding the leaves and webbing the delicate shoots will inflict injury. Larvae skeletonise leaves by weaving webs among them and feeding on soft tissues from the inside. Fully grown caterpillars gorge on delicate foliage. Stunting results from the preference for feeding on the apical tips. Egg laying also results in the destruction of apical branches. Leaf quality and yield are negatively impacted.

Management

- Flood irrigation.
- Release of pupal parasitoid, *Tetrastichus howardi* @ 20,000 / ac
- Release of egg parasitoid, *Trichogramma chilonis* @ 2cc / ac 10 days after pruning.
- Spraying of dichlorvos @ 1ml /l 30 days after pruning.
- Mechanical clipping and burning of affected shoot 40 days after pruning.

4. Thrips

Damage the leaves' and sap's epidermal tissues, early maturity, moisture loss, and a decrease in total sugars and crude protein. Too little leaves to raise silkworms.

Fig. 6.4: Thrips

Signs and symptoms

Leaf streaks are seen in the early stages of the infestation. The leaves mature to a yellowish brown colour in advanced stages.

Management

- 2 ml/lit spray of dichlorvos 76 WSC.

5. Termites and White Ants

Damage is done by colony workers. Termites create an earthen covering for the stem, consume the bark, and ultimately destroy the plant, becoming dry and producing fewer leaves. Dry twig mulching promotes population growth in endemic areas. Sandy and red loam soils show more damage than others.

Fig. 6.5: Termites

Fig. 6.6: White ants

Management

- Locating termite colonies and eliminating them by eliminating the queen termite

- Mounds are treated with 50 ml of chlorpyriphos 20 EC.
- Applying 50 ml of chlorpyriphos 20 EC via swabbing or soaking to the base of established plants.

6.3 Mulberry Diseases

1. Powdery Mildew

Phyllactinia corylea is a fungal pathogen that causes powdery mildew mulberry disease. The emergence of white, powdery spots on the lower surface of the leaves is the main sign of this illness. In cases of severe illness, the white, powdery spots become brownish-black, and the leaves become coarse, yellow, and lose their nutritional value. Even after being gently crumpled by hand, the heavily affected leaves would turn into brittle powder (**Fig. 6.7**). Crop loss might reach 10% to 15%. The illness is most common from August through January, with its peak intensity occurring in October. Higher relative humidity and lower temperature promote the growth of illness. Powdery mildew is more common in untended gardens.

Fig. 6.7: Powdery mildew

Management

Spraying 0.2 % of Carbendazim or 0.1-0.2 % Captan or 0.05 % Dinocap fungicide is preferable. The fungicides solutions should be sprayed to lower portion of the leaves. 4.5 kg/acre of "Seri-Mildewguard" must be sprayed. 2.5 kg of mildew guard should be dissolved in 125 litres of water and sprayed followed by second spray of 2.0 kg of mildewguard in 100 litre spray after the 10 days of first spray.

Leaves should not be used as feed for silkworms for 15 days if Carbendazim or Captan sprayed and 20 days, if sprayed a Dinocap fungicide.

2. Leaf Spot

The fungus ***Cercospora moricola*** is the cause of mulberry leaf spot disease. The mulberry leaf surface will initially show little, uneven, light brown dots as the illness progresses. Subsequently, the brownish spot gets larger and eventually joins with the surrounding yellow patches to form a distinctive "shot hole" before dying off (**Fig.6.8**). Leaf spot disease is thought to be responsible for 10% to 20% of leaf loss. This disease usually affects leaves that are 18 days old. Leaf spot disease occurs during rainy season, particularly June to October. It spreads through rain water drops, so its intensity becomes low once rainy season is over.

Fig. 6.8: Leaf spot

Management

Since it spreads through a rain water droplet, it is better to remove all old or diseased leaves to avoid infection.

It can be controlled by spraying a fungicide, 0.2 % of Carbendazim or Benomyl (2 g in 1 litre of water) for two times at the interval 15 Days. Leaves should not be used as silkworm feed for 15 days if fungicide sprayed.

3. The Leaf Rust

The ***Peridiospora mori*** fungus is the cause of mulberry leaf rust disease. The first signs of rust sickness are round, pin-head-sized light brown spots that, as the illness progresses, turn into dark brown spots. There can be up to 5% leaf loss. Regardless of when the shoots were pruned, the severity of the rust greatly increased with increasing shoot age (**Fig. 6.9**).

Although this condition occurs from October to February, its highest intensity is noted from November to January; still, it is more prevalent in the wet and winter seasons.

Fig. 6.9: Leaf rust

Management

Harvesting a mulberry leaf at the right time minimized the occurrence of leaf rust disease. Wider spacing during mulberry plantation will be better option to avoid the rust disease. If need comes or the disease severity is more, spraying a 0.2 % Carbendazim is better management option. Leaves should not be used as silkworm feed for 15 days, if fungicide sprayed.

4. Canker in Mulberry Stem

The ***Lasiodiplodia theobromae*** is the causative agent of mulberry stem canker. The illness causes mulberry cuttings to not sprout and increases nursery mortality. This disease is annual in nature, with a fatality rate ranging from 30% to 70%. Beneath the soil's surface, the bark of the stem is rotting and flaking. There are black eruptions on the bark of the affected stem section and black mycelial threads visible beneath the diseased bark. When these plants are closely inspected, the stems show signs of drying out and discolouration. In the later stages of the disease, there is a lot of dark structure in clusters and a greenish-black lesion on the stem.

Fig. 6.10: Stem canker

Management

The infected cuttings from the nursery should be removed immediately. The mulberry cuttings must be immersed in 0.2 % Carbendazim for 30 minutes. The fungicide should be sprayed once again after 15 days of nursery. If the disease intensity is more in nursery bed, the soil should be treated with 8 kg of Carbendazim or Captan 15 days before planting mulberry cuttings in a nursery. Immediately after harvesting mature plants, the cut portion of the plants should be sprayed with 0.2 % of Carbendazim/Captan fungicide solution.

5. Mulberry Root Rot Disease

Fig. 6.11: Root rot

The causal organism for mulberry root rot disease is ***Lasiodiplodia theobromae***. The V1 and S36 mulberry varieties are more prone to this disease. This illness strikes at any time of the year. On the stem close to the point where the root and soil meet, the infected plant exhibits a distinctive rotting symptom. The mulberry plant eventually disappears. A black fungus will grow between the plant's cortex and stem. Peeling off the cortical section is simple. There is a greater likelihood of fungal pathogens attacking the plant's harvested shoot section. (**Fig. 6.11**)

Management

The infected plants must be uprooted and burnt. Immediately after harvesting mature plants, the cut portion of the plants should be sprayed with 0.2 % of Carbendazim/Captan fungicide solution. 4 kg of Seri-Bioguard should be added to mulberry garden per year/acre in two equal splits.

6. Root Knot

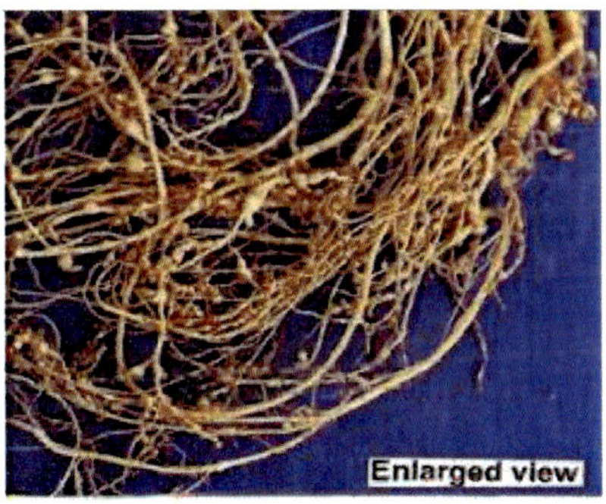

Fig. 6.12: Root knot

Meloidogyne incognita is a nematode that causes root knot disease. It happens all year long, with sandy soil experiencing increased intensity. In tiny plants, disease incidence is readily discernible. The seriously impacted mulberry plants displayed slowed development and leaf margin yellowing. Even when the soil has enough moisture content, the leaves nevertheless exhibit signs of dryness. On roots are thousands of circular knots. Consequently, the efficiency of the root absorption capacity is lowered and hindered by the formation of root knots. The older the plant gets, the worse the sickness gets (**Fig. 6.12**).

Management

The nematode infested soil should be deep ploughed and exposed to sunlight. The disease free plants should be planted in a new garden. Neem or pongamia cake should be added at 2 tons/hectare in 4 equal splits. The infestation may be avoided by growing chrysanthemum, sesame or jute. Apply @ 40 kg of Carbofuran (Furadan 3g)/hectare in 4 equal split around the plant. 4 kg of Seri-Nematoguard should be added to mulberry garden per year/acre in two equal splits. Leaves should not be used as silkworm feed till 45 days if Carbofuran (Furadan 3g) applied.

References

Mulberry Diseases & Microbes - Karnataka state sericulture research and Development Institute. https://kssrdi.karnataka.gov.in/page/Contributions/Moriculture%20Technologies/Mulberry+Diseases+&+Microbes Sericulture::Pest Management

Outcomes Assessment

Part A

Answer the following questions true or false. [1×5=5 Marks]

1. In leaf webber adults defoliate the mulberry plant. [True/False]
2. Root rot disease is caused by a nematode. [True/False]
3. Leaf spot is caused by *Cercospora moricola.* [True/False]
4. Powdery mildew is caused by *Peridiospora.* [True/False]
5. For the treatment of powdery mildew infected plants are uprooted and burnt. [True/False]

Part B

Answer the following questions. [2×3=6 Marks]

1. Write about the symptoms of damage caused by Papaya mealybug and its management.
2. Write about the symptoms of damage caused by Thrips and its management.
3. Write about the symptoms of damage caused by canker on the mulberry stems and its management.

Part C

Write a brief note on each of the following. [5×2= 10 Marks]

1. Write about the symptoms of damage caused by diseases on the mulberry plant and their management.

2. Write about the symptoms of damage caused by insect pests on the mulberry plant and their management.

Part D

Select the correct answer from the four choices given under each question.

[1×5=5 Marks]

1. Which nematode species infects mulberry plant?
   a. *Meloidogyne sps.*
   b. *Heterodera sps.*
   c. *Longidorus sps.*
   d. *Cacopaurus sps.*
2. *Lasiodiplodia theobromae* causes which disease in mulberry plant?
   a. Powdery mildew
   b. Root Rot
   c. Canker
   d. Anthracnose
3. *Peridiospora mori* causes which disease in mulberry?
   a. Root Rot
   b. Canker
   c. Leaf rust
   d. Downy Mildew
4. Which of the following insect causes apical shoot malformation?
   a. Papaya Mealy Bug
   b. Thrips
   c. Termites
   d. The pink mealy bug
5. Seri-Bioguard is used for the control of which disease?
   a. Leaf Rust
   b. Root Rot
   c. Canker
   d. Powdery mildew

Answer Key: 1. **a**, 2. **b**, 3. **c**, 4. **d**, 5. **b**.

7

Identification of Cocoons of Important Breeds

7.1 Introduction

The three factors used to categorise silkworm races are: (1) origin, (2) voltinism and (3) moulting. In addition, they are categorised according to their shape and colour. For more than 2,000 years, people have tamed and raised the silkworm *Bombyx mori*. It is a monophagous insect that exclusively consumes the leaves of mulberries. As a result, mulberry silk is typically meant when the word "silk" is used.

7.2 Classification Based on the Place of Origin

(a) **Indian Races:** Native to South East Asia and India; these races are indigenous. They have a longer larval stage and can withstand high humidity and temperatures. The larvae and cocoon are modest in size. The cocoon is often spindle-shaped and coloured in shades of green, yellow, or golden yellow. With a lower shell percentage, the cocoon shell is thin and most of them are multivoltine.

(b) **Japanese Races:** Japan's indigenous races comprise this group. The larvae are strong. The cocoon resembles a peanut. The majority of cocoons are white, while some are yellow in hue. There are more double cocoons in the ratio. The silk strand is thick and short and of poor quality and are bivoltine or univoltine.

(c) **Chinese Races:** These ethnic groups are native to China. The larvae are resistant to high temperatures; however, they are not resistant to high humidity. The larvae are simple and energetic. They develop swiftly and consume mulberry leaves with a voracity. The morphology of the cocoon is spindle-shaped in rare instances and elliptical or spherical in most. The cocoon has a golden, white hue. The reelability is good, and the cocoon filament is fine. They are mostly univoltine, bivoltine and multivoltine.

(d) European Races: Native to Europe and Central Asia are these races. Their larvae live for a long time and aggressively consume mulberry leaves. Larvae are not strong enough to withstand extreme heat or humidity. The cocoon is large and slightly constricted. The cocoon has a white or cream hue. The filament's length of the filament is more. The ratio of double cocoon is low. Reelability is good.

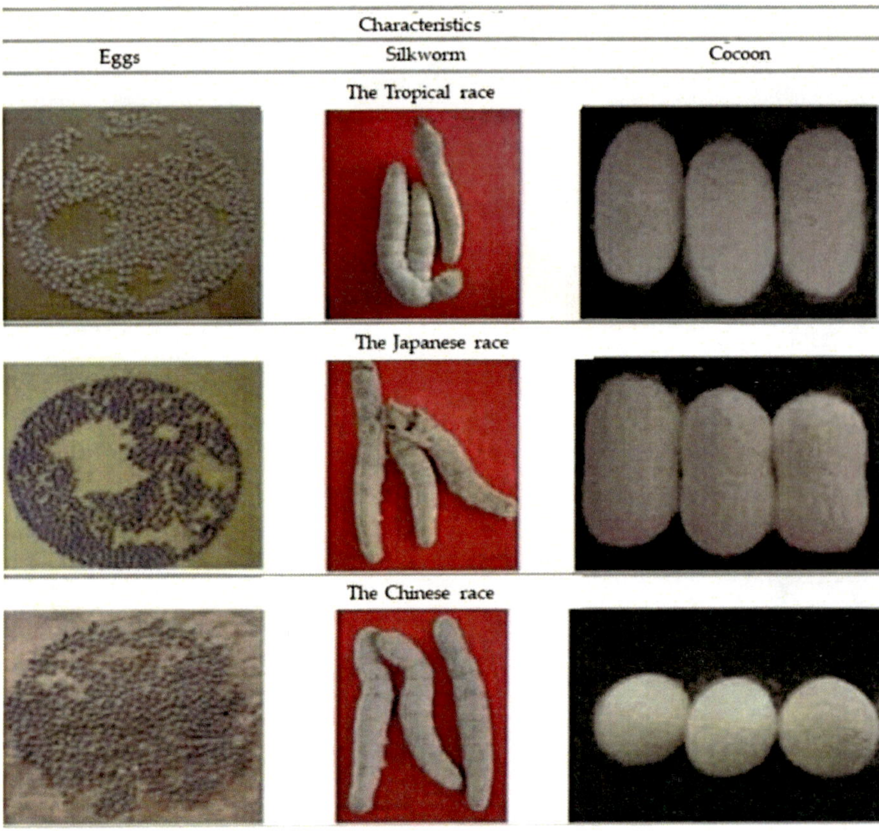

Fig. 7.1: Races of silkworms

7.3 Classification based on Voltinism

Voltinism is the number of generations an organism undergoes in a year. Silkworms are divided into three groups according to their voltinism: (a) Univoltines, (b) Bivoltines, and (c) Multi or Polyvoltines.

(a) Univoltines: Every year, these races produce just one generation. The body of the larva is big. The weight of the cocoon, shell, ratio, and filament of the cocoon are all high. The quality of the cocoon filament is good.

(a) **Bivoltines:** Every year, these races produce two generations. The silkworms are stronger and more unified. In comparison to uni-voltines, the cocoon weight, cocoon shell weight, shell ratio, and cocoon filament weight are all lower than those of multi-voltines. When compared to univoltines, larvae are resilient.

(b) **Multivoltines:** They have a brief lifespan. They live in a multigenerational. They have many generations in a year. Larvae are robust and can withstand high temperature. The cocoon size is small. The cocoon weight, shell weight, shell ratio and cocoon filament weight are less than bivoltines. Cocoon filament is fine.

7.4 Classification Based on the Moulting of Silkworms

Moulting is the process by which a larva sheds its skin as it grows and progresses through its life stages. Silkworms are divided into four groups according to their moulting habits: (a) tri moulters; (b) tetra moulters; (c) penta moulters; and, very infrequently, bi and hexa moulters. Tetra moulters are mostly raised for market.

7.5 Classification Based on the Shape and Colour of Cocoons

Colour: Distinct silkworm races have distinct cocoon colours. The cocoon's colouration is only a result of **sericin**, and after degumming, it turns white. The races of silkworms determine the pigments' kind and distribution. The outer and interior layers of the shell can have different colours. The influence of temperature and humidity during rearing affects the colour's tint.

Cocoon's Form: A further racial characteristic of the cocoon is its **shape**, which is somewhat determined by the kind of mountages employed. Spindle, spherical, and restricted elliptical are the three shapes that make up a cocoon.

Cocoons that have extremely pointed ends or a severe constriction in the middle are not good for reeling because they break more easily, which reduces reelability and raises waste percentage. During the brushing procedure, pointed cocoons often acquire holes that render them unusable for additional reeling. For reeling, the oval and slightly constricted cocoons work well. The majority of the long, oval cocoons produced in Karnataka are bivoltine, or *C. nichi*, (a male parent) and pure Mysore (a female parent) hybrids. The silkworm race determines the cocoon's volume and size. The unit of measurement is cocoons per litre. In uni- or bivoltine cocoons, there are roughly 90–120 cocoons per litre. It is considerably more with multivoltine races.

7.6 Mulberry Silk

The mulberry silk caterpillar, belonging to the order of Lepidoptera, genus ***Bombyx,*** which can be cultivated interior, produces over 90 percentage of the sector output of raw silk used commercially. There are other forms of wild silkworms under the Family Saturniidae. Silk is crystalline, homogenous in structure, hygroscopic in nature, mild in weight and is the longest and the strongest of all natural fibres, soft lustrous and hygienic, silk also has an extraordinary affinity to dyes. Silk does not seize hearth as speedy as nylon and wool. It is far elastic and has an elongation of 20 percentage. Its breaking strength is set four grams in line with denier. Mulberry, the fundamental diet of the silk caterpillar claims almost 60 percentage in silk production value. An oz of silkworm eggs yields 25-30 kg of cocoon; the metamorphic procedure from egg to larvae to cocoon, needs a ton of mulberry leaves. A mulberry land hectare can produce 15-50 MT of foliage depending on agro-climatic situations and usage of nutrients. In comparison to other agricultural vegetation, sericulture is a particularly labour-intensive profession due to the tasks involved in preparing a silkworm weight loss plan through cultivation, harvesting, slicing, and reducing leaves. This is true even with mechanization and improved home equipment. The only nation in the world that can claim three distinct types of non-mulberry silks—Tasar, Eri, and Muga—is India. The production of uncooked silk accounts for 9.5% of total silk production. The state benefits from non-mulberry production since it provides for the socioeconomic needs of the tens of millions of tribal people who live in rural India.

7.7 Tasar Silk and Eri Silk

The famous handspun tussah so famous in the US and Europe, comes from the obscure remote forest regions of 5 Indian States —Bihar, Madhya Pradesh, Odisha, Andhra Pradesh and Uttar Pradesh. The tropical Tasar Silkworm is called the ***Antheraea mylitta*** and the temperate Tasar Silkworm is called the ***Antheraea proylei***. They are univoltine, bivoltine or tetravoltine. The wild tasar silk cocoon putting on a peduncle from a spring is unsuitable as a fruit. The cocoons are visible in numerous hues of brown, gray, yellow and crimson.

The Tasar cocoon unlike other cocoon has three layers of continuous filaments, plus the internal finer strands of the chrysalids. The first two layers are precise in that the outer layer is solidly cemented into a tough shell. It measures approximately 5 × 3 cm. The continuous internal filament is lustrous, and varies in colour from brown to various sunglasses of tan to silver grey. It is far hard to reel due to the fact the filaments do not readily form a thread. The cocoons need to be degummed or boiled for approximately 6 hrs in a crude solution of soda and ash and then they are semi-dried before they are taken up for reeling. The reeling is carried out on an easy hand spindle known as Natwa. A reeler can reel about 80 cocoons consistently within a day. The tribals have also developed techniques of indigenous dyes made from herbs and fruits. However it is undyed yarn, which is greater in demand for export weaving.

The second sort of non-mulberry silk is known as eri silk. The eri silk worm referred to as ***Philosamia ricini*** and the other species ***Philosamia cynthia*** are very famous for eri silks. The term "eri" in Assam is derived from the castor oil plant referred to as Varanda or Arandi on which the silkworm feeds. Castor plants (*Ricinus communis*), which grow alongside the Brahmaputra, are the main food source for silkworms. They are also grown as area flora in other places, where the leaves are used to raise eri worms and the seeds are used to extract oil. It is a perennial plant. The other food vegetations are Keseru (*Heteropanax fragrans*), Cassava (*Manihot utilissima*), Papaya (*Carica papaya*), Champa (*Plumeria acutifolia*) etc. The worms that ate up Keseru leaves spin smaller and more compact cocoons than those that ate up castor leaves. For the duration of the early stages of raising, the rearers preserve money or choose the alternative plant life that works best in an emergency. The life cycle of the eri worm from egg to egg takes approximately 6 weeks in summer and 12 weeks in wintry weather. During the early phases of raising, the rearers either save money or, in case of an emergency, select the most effective substitute plant life. Such cocoons are the clearest and of high quality. But some rearers decide to expose their cocoons to the sun which allows to stifle the pupae. Cleared cocoons weigh approximately 2700 per kg. The bulk of the cocoons are spun into yarn and woven into cloth by the rearers for their very own use. The yarn is dyed and woven on handloom like some other thread. Assam eri silk is popular for softness, durability and handcrafted in harmony with nature, with roots in tradition for many countries. The fabric possesses a natural cream shade and an extraordinary herbal hand-woven appearance. The twisting and spinning processes are primitive. However, the cloth keeps the herbal beauty of the fibres and makes an ideal cloth for style designers. Assam (Jagi avenue) and Bihar (Bhagalpur) spun silk by mills set up by the governments and make use of maximum of the cocoons and different silk waste.

7.8 Muga Silk

Notwithstanding the upsurge of mill-made material, the Assamese, even now-a-days proudly put on native outerwear and turbans made from muga silk. In Assam, ladies like to put on silk wraps, Mekhelas embroidered with arresting designs. Assam produces nearly 60-70 MT of muga raw silk each year. *Antheraea assama* is a semi-domesticated muga silkworm that produces multiple generations annually. The remaining 25–30% of the cocoon thread is hollow (clean), and the remaining 70–75% is strong . Because of this, it is easy to dye and absorbs a lot of oxygen, salts, acids, and alkalies. Because it is porous, silk absorbs water molecules both when it is submerged in water and from its surroundings. While the burden increases by 30–35%, the fabric's diameter swells by 16–18%. One essential feature of clothing materials is their robust capacity to absorb moisture and water. A lot of mechanical and physical homes are heavily traded since the silk fabric absorbs water and humidity. The water content requirement rate has remained steady at 11%.

References

Central Sericultural Research and Training Institute (2005). Annual report 2004-2005.
Government of Assam (n.d). Mulberry silkworm moth. Sericulture Department, Assam.
Karnataka State Sericulture Research and Development Institute. (n.d)

Outcomes Assessment

Part A

Answer the following questions true or false. [1×5=5 Marks]

1. Mostly Indian races are univoltine. [True/False]
2. Mostly Japanese races are multivoltine. [True/False]
3. Indian cocoon represents peanut shape. [True/False]
4. Cocoon size is large in multivoltines. [True/False]
5. Colouration of cocoons is due to sericin. [True/False]

Part B

Answer the following questions. [2×4=8 Marks]

1. Write about the classification of cocoons based on voltinism.
2. Write about the classification of cocoons based on place of origin.
3. Write about the classification of cocoons based on colour and shape.
4. Write about the classification of cocoons based on moulting.

Part C

Write a brief note on each of the following. [5×2=10 Marks]

1. Write about the identifying characterstics of mulberry, eri, muga and tassar silkworm cocoons.
2. Write about the different types of classifications of cocoons.

Part D

Select the correct answer from the four choices given under each question. [1×5=5 Marks]

1. An oz of mulberry eggs yields about

   a. 20 kg cocoon b. 25-30 kg cocoon

   c. 40-45 kg cocoon d. 50 kg cocoon

2. The races which produces multiple generation in a year

   a. Univoltine b. Bivoltine

   c. Multivoltine d. None of the above

3. The colour of European race cocoon is

   a. Green b. Yellow

   c. Creamy white d. Red

4. Muga silkworms are grown in which state of India ?

   a. Assam b. Odisha

   c. Kerala d. Sikkim

5. In univoltine/ bivoltine how many cocoons are produced per liter ?

   a. 50-60 b. 20

   c. 130 d. 90-120

Answer Key: 1. **b**, 2. **c**, 3. **c**, 4. **a**, 5. **d**.

8

External Morphology of Life Stages of *Bombyx mori* L.

8.1 Life Cycle

As in the case of a typical Lepidopteran insect, the silkworm passes through four distinct stages, i.e., egg, larva, pupa and adult during its life cycle, the duration of which may last for six to eight weeks depending upon racial characteristics and climatic conditions. Multivoltine races found in tropical areas have the shortest life cycle with the egg, larval, pupal and adult stages lasting for 9-12 days, 20-24 days, 10-12 days and 3-6 days respectively. (**Fig.8.1**)

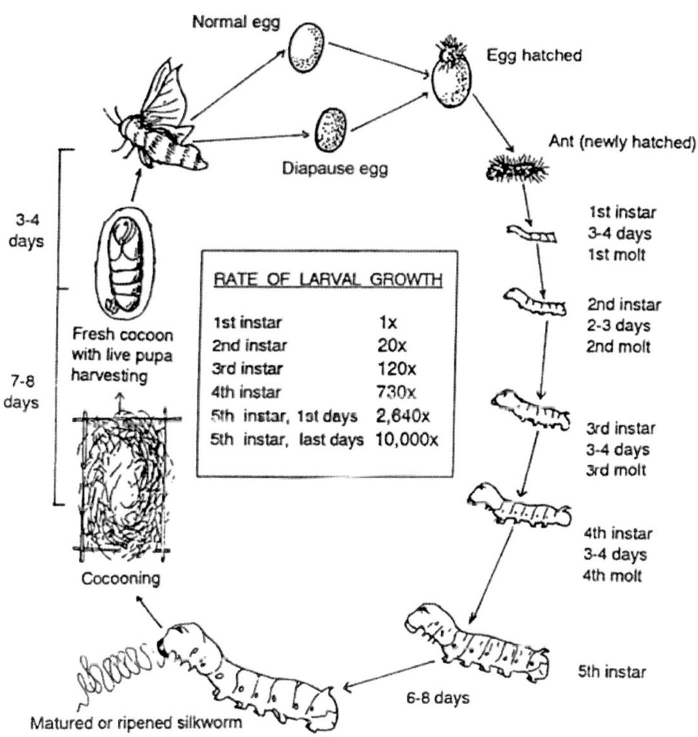

Fig. 8.1: Life cycle of silkworm

In uni/bivoltine races, the egg period of the activated egg may last for 11-14 days, the larval period 24-28 days, the pupal period 12-15 days and the adult stage 6-10 days. In nature the univoltine races produce only one generation during the spring, and the second generation egg goes through a period of rest or hibernation till the next spring. In the case of the bivoltine races, however, the second generation egg does not hibernate and hatches within 11-12 days and produces the second generation normally during summer. But it is the third generation egg which undergoes hibernation and hatches only the next spring, thus producing only two generations in a year (bivoltine). In multivoltine races, the life cycle is the shortest because of the warmer ecological conditions where they are reared, and so they may yield as many as seven to eight generations in a year in tropical sericultural areas such as India, Thailand, etc. Silkworm rearing is, therefore, continuous in tropical areas whereas in sub-tropical and temperate zones it is mostly seasonal lasting from spring to early autumn.

8.2 Larval Life

In sericulture it is the larval life that is of direct importance to the rearer since he has to take care of the worms very carefully (**Fig.8.2**). The larval life may last from 20-24 days in the case of the multivoltine species in tropical areas or 24-28 days in the case of uni and bivoltine races in temperate areas, being shorter under warmer summer and autumn conditions, and somewhat longer under cooler spring conditions. During the larval life the worm moults or casts off its skin four times to be able so as to grow. There are also varieties of worms which moult three and five times but they are not of importance economically. In view of the four intervening moults, the larval life is divided into five distinct stages or instars which are referred to popularly as five different ages. The first three instars are referred to as young ages and the fourth and fifth instars as late ages. The duration of the different instars and moulting periods for the different races is as follows (**Table 8.1**):

Table 8.1: Duration of different instars and moulting periods

Duration of different instars and moulting periods					
	Multivoltine races			Uni and bivoltine races	
		Duration	**Temperature and humidity conditions**	**Duration**	**Temperature and humidity conditions**
1	instar	3 days	27°C 80-85% R.H.	3 days	27°C and 85% R.H.
2	moult	20 hours		20 hours	
3	instar	2 days		2 days	
4	moult	20 hours		20 hours	
5	instar	3 days		3 days	25°C and 30% R.H.
6	moult	1 day		1 day	
7	instar	4 days	25-26°C 70-80% R.H.	5 days	22-24°Cand 75% R.H.
8	moult	1 day		1 day	
9	instar	6-7 days		9-10 days	20-23°Cand 70%
Total Duration		22-23 days		26-27 days	

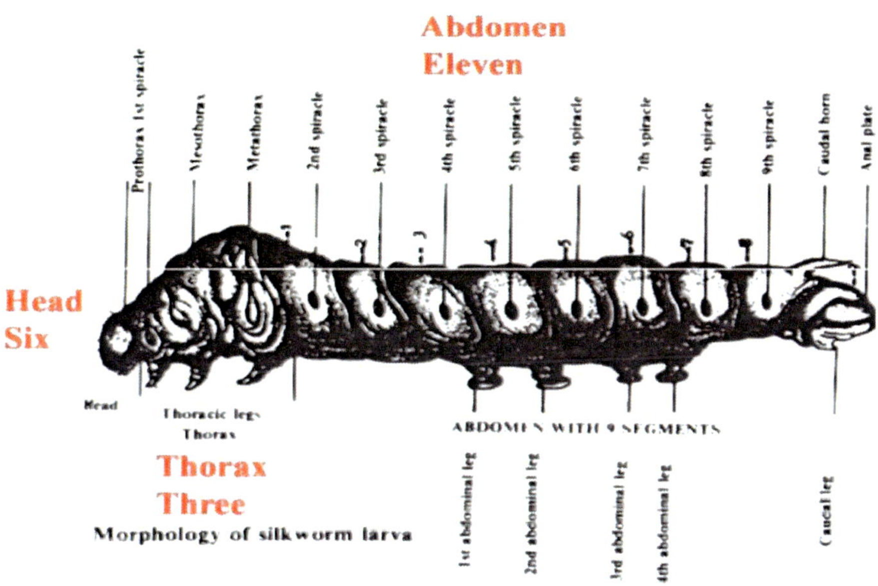

Fig. 8.2: Morphology of silk worm larva

8.3 Moulting

Each larval instar can be broadly divided into two phases i.e., the feeding phase and the moulting phase. After feeding voraciously and having attained full growth for the particular instar the worm loses its appetite and the larva prepares to moult and cast off its old skin. Prior to each moulting, the larva stops feeding and rests with its head held up. The lustrous body skin gradually becomes translucent, loose and wrinkled and the worm becomes dull in appearance and wanders about in search of a resting place. It emits a silky substance with which it fixes itself to dry leaves. The moulting periods last for 20 hours being shortest in the second moult, followed by the first, third and fourth. This resting for moulting is often referred to as going to sleep and the coming out of the worm from moult as waking up. During the moulting period the worms should not be disturbed so that the process of moulting is not interfered with and uniform moulting is ensured.

8.4 Mature Worms

After the silkworm passes through four moults, it reaches the fifth and final instar when it attains its maximum weight a day prior to maturity and before it stops feeding. At its maximum weight it is about 10000 times its own weight at the time of hatching and this phenomenal growth takes place within the short span of 20-25 days. When the worm is fully mature for mounting, it loses it appetite, stops feeding, and excretes soft faeces with high moisture content. At this stage because of its voluminous increase in size, the silk gland is visible through the body integument and, therefore, the thorax and body segments of the mature larva appear translucent. In fact the silk glands are so enlarged as to account for nearly 40 per cent of the body itself. This is characteristic of the ripened worm and serves as guidance for picking the mature worms for mounting. The mature worms become very restless and raise their heads in search of support so as to be able to start spinning.

8.5 Spinning of the Cocoon

The spinning of the cocoon starts almost immediately after mounting and in 48-72 hours spinning is completed by the mature worm. In another day or two the worm transforms itself into the pupa within the cocoon.

The pupal period may last for **8-14 days** after which the adult moth emerges slitting through the pupal skin, and piercing the fibrous cocoon shell with the aid of the alkaline secretion (**cocoonase- a protease enzyme**) that softens the tough cocoon shell. The adult moths are ready to copulate immediately after emerging from the pupa and the female then lays the eggs. Adult life is short, lasting from 3-10 days depending upon races and seasons. The adults do not feed and are also incapable of flight in size and generally sluggish while the males are somewhat smaller and more active. A female of the multivoltine variety may lay on an average approximately 400 eggs while the bivoltine varieties of silkworm moths lay about 500 eggs.

Silkworms pass through a complete metamorphosis intermediate stage of larva (caterpillar) and pupa (cocoon). The morphology of the *Bombyx mori* has different stages in its entire life cycle. The size, weight, shape, colour of the egg as well as number per laying vary among different strains. The silkworm eggs are tiny and weight of a single egg is about 0.55 to 0.6 mg. The silkworm eggs measure 1 to 1.3 mm, flat, ellipsoid or oval and flat at the ventral side, micropyle at the anterior pole and slightly off. The respiratory canals are distributed in large numbers over the entire surface of the egg and these are funnel shaped. The air necessary for respiration enters through a special character. Races producing white cocoons lay pale yellow eggs, do not feed and are also incapable of flight due to domestication over thousands of years. The females are larger in size and generally sluggish while the males are somewhat smaller and more active. A female of the univoltine variety may lay on an average approximately 400 eggs while the average number for the bivoltine varieties of silkworm moths is from 500-600.

8.6 Morphology *Bombyx mori*

It consists of complete metamorphosis (Holometabola) from egg to the adult stage through two intermediate stages of larva (caterpillar) and pupa (cocoon). Its different life-cycle stages are discussed below:

8.6.1 Egg

The size, weight, shape, colour of the egg as well as number per laying vary among different strains. The weight of a single egg is about 0.55 to 0.6 mg, so that there are 2000 eggs per 0.115 kg. The silkworm eggs measure 1 to 1.3 mm in length and 0.9 to 1.2 mm in width.

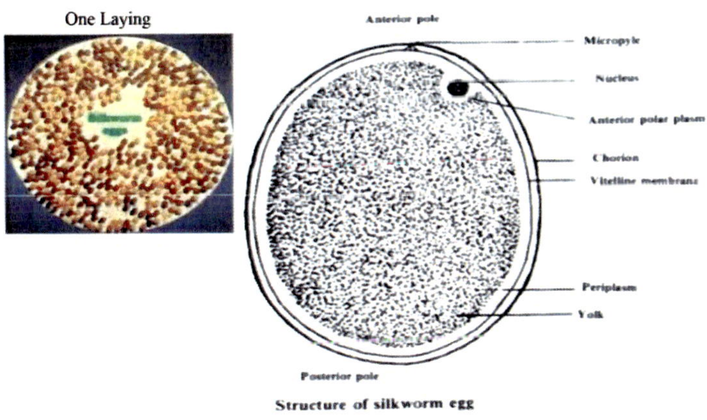

Fig. 8.3: Structure of silk worm egg

The protective covering of the egg is chorion, a small opening micropyle at the anterior pole and slightly off-centre. The sperm enters through this opening (**Fig.8.3**). The respiratory canals are distributed in large numbers over the entire surface of the egg and these are funnel shaped. The air necessary for respiration enters through these respiratory canals. The colour of the egg is pale yellow. The females are larger in size and generally sluggish while the males are somewhat smaller and more active. The sperm enters through this opening for fertilization. The respiratory canals are distributed in large numbers over the entire surface of the egg and these are funnel shaped. The colour of the egg is deep yellow. In both cases the Japanese races lay slightly darker eggs than the Chinese races. In the case of the hibernating eggs laid by bivoltine and univoltine races, the egg colour changes and becomes dark brown or purple with the deepening on the colour of the serosal pigments.

8.6.2 Larva

The silkworm larva when newly hatched is black or dark brown. It has a large head and the body is densely covered with bristles so that it looks like a hairy caterpillar. There are four pairs of tubercles; the subdorsal, supraspiracular, infraspiracular and basal tubercle each of which carries three to six setae. As the larva grows, it becomes smoother and lighter in colour due to the rapid stretching of the cuticular skin during the different instars of the larval stage. The larval body is composed of the same segments as any insect: head, thorax and abdomen. The entire body is covered with a thin and elastic chitinous cuticle which is capable of being extended considerably to permit rapid growth of the larva during any instar. The integument consists of the epidermis and the cuticle over it (**Fig.8.4**).

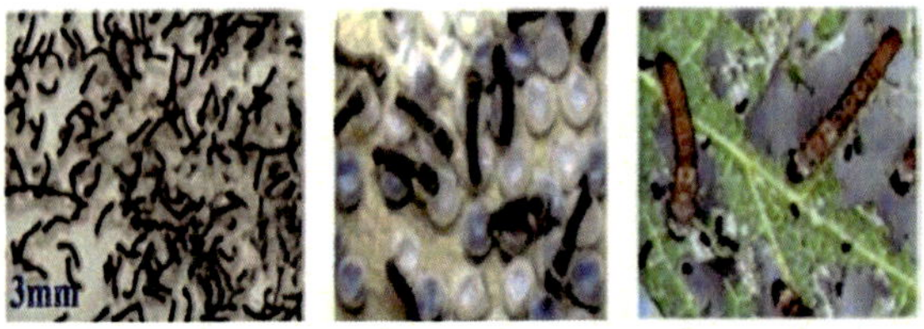

Newly hatched larvae Enlarged Second Instar larvae

Third Instar Larvae Fifth Instar Larvae

Fig. 8.4: Instars of silkworm larvae

8.6.2.1 Head

The head is made up of a skull and six body parts that have been fused together. Appendages that have been changed to become antennae, mandibles, maxillae, and labium are carried by the second, fourth, fifth, and sixth segments, respectively. The median epicranial suture is prominent and well-developed. Likewise, the labrum and clypeus are noticeable on the outside. Six pairs of ocelli, or larval eyes, are situated behind and slightly above the antennae's base. The two antennal segments serve as the body's sensory organs, or feelers. The mandibles are strong, well-developed, and designed for mastication. On the palpi's ventral side are the maxillae. The maxillary palpi and maxillary lobe distinguish between different food tastes. A large, somewhat chitinized mentum is carried ventrally by the labium. The prementum is black and chitinized. A median process from the silk bave or thread is carried distally by the prementum. The spinneret has sensitive labial palpi on both sides. Enhanced The silkworm's head A silkworm's head in a colored scanning electron microscopy (SEM). The prothoracic segments, which are the three body segments that make up the thorax, carry a pair of legs ventrally, each of which is made up of three jointed segments. Conical in shape and equipped with strong distal claws, they are the actual legs. The labrum and clypeus are similarly noticeable, but these claws are not employed. Six pairs of ocelli, or larval eyes, are situated behind and slightly above the antennae's base. There are two antennae, which are sensory organs (feelers) made up of five joined segments. The mandibles are strong and well-developed. The maxillary lobe, palpi, stipes, and cardo make up the maxillae on the ventral side of the mouth. The maxillary palpi and maxillary lobe distinguish between different food tastes. The ventrally large, somewhat chitinized mentum contains the labium. The prementum is black and chitinized. Silk is discharged from the silk gland to produce the silk bave or thread via a median process or spinneret that is carried distally by the prementum. The spinneret has sensitive labial palpi on both sides. (**Fig.8.5**)

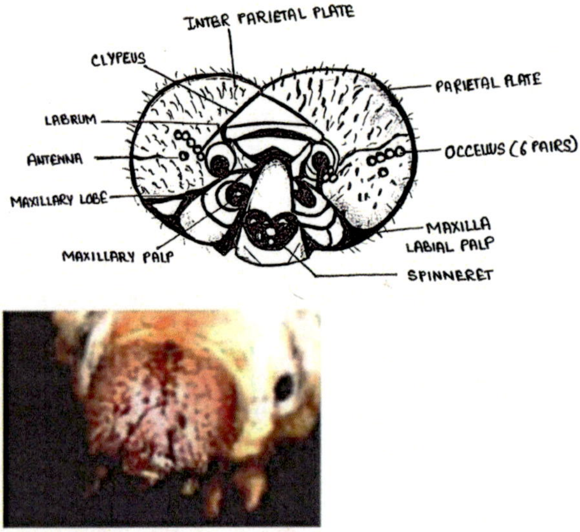

Fig. 8.5: Head of larva

8.6.2.2 Thorax

The thorax consists of three body segments called the pro, meso and metathoracic. Thoracic segments carry ventrally a pair of legs each comprising in turn three jointed segments. These are the true legs which are conical in shape and carry sharp distal claws. These claws are not used for crawling but for holding mulberry leaves while feeding. All the silkworm larvae that show body larval markings carry the so-called eye spot on the dorsal side of the mesothorax.

8.6.2.3 Abdomen

The abdomen is comprised of eleven body segments although only nine can be distinguished as the last three are fused together to form the apparent ninth segment, the anal plate and the caudal legs. The third to sixth and the last abdominal segments each bear a pair of abdominal legs which are fleshy, unjointed muscular protuberances. At the extremity they form a sort of disc with a series of hooks inwardly curved and arranged in a semicircle. On the dorsal side of the eighth abdominal segment, the larva carries the caudal horn. The abdominal segments carry the sexual markings which develop distinctly in the fourth and fifth instars in the eighth and ninth segments on the ventral side. In the female the sexual markings appear as a pair of milky white spots in each of the eighth and ninth segments. The pair of spots on the eight segments is known as Ishiwata's fore glands and the pair on the ninth segment is referred as Ishiwata's hind glands. In the male a small milky white body (Herold's gland) appears at the centre of the ventral side between the eighth and ninth segments. There are nine pairs of spiracles placed laterally on either side of the body. They are found on the first thoracic segment and the first eight abdominal segments. These are the breathing or respiration pores. The larval skin or integument consists of the cuticle and the hypodermis. The cuticle is made of chitin as well as protein and is covered with a thin layer of wax. Nodules are found all over the body surface of the silkworm larva. The distribution of these nodules differs according to the variety of silkworms, being less in number in varieties having lustrous body surfaces, such as the Chinese bivoltine and the multivoltine varieties. The nodules are found in larger numbers over parts of the body where normally the larval markings are found. The larval markings in silkworms are caused by skin pigment.

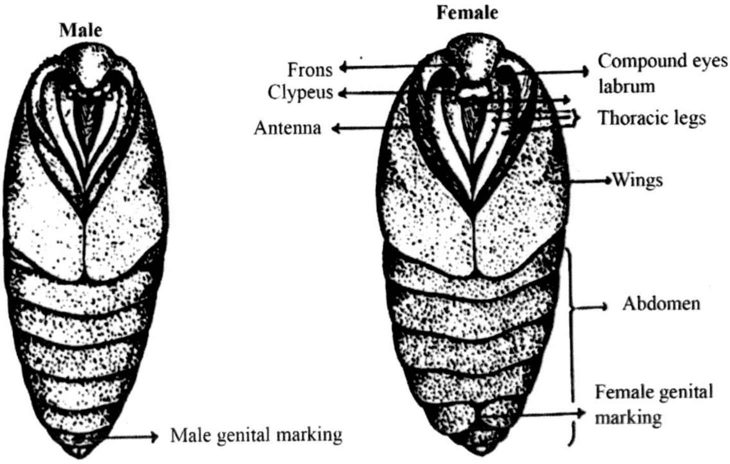

Fig. 8.6: Male and female pupa of *Bombyx mori*

8.6.3 Pupa

The pupal stage is generally called the resting, inactive stage of the silkworm when it is incapable of feeding and appears quiescent. This is a misnomer. The pupal stage is a transitional phase during which definite changes take place. During this period of biological activity, the larval body and its internal organs undergo a complete change and assume the new form of the adult moth. The mature silkworm larva passes through a short transitory stage of pre-pupa before becoming a pupa. During the pre-pupal stage, the dissolution of the larval organs takes place and this is followed by the formation of adult organs during the pupal stage. Soon after pupation the pupa is white in colour and soft but gradually turns brown to dark brown and the pupal skin becomes harder. The prominent morphological parts visible are a pair of large compound eyes, a pair of large antennae, fore-and hind-wings and the legs. Ten of the abdominal segments can be seen on the ventral side, but only nine of them are visible on the dorsal side. Seven pairs of spiracles can also be discerned in the abdomen; the last pair is non-functional. Sex markings are prominent and it is much easier to determine sex in the pupal stage than in the larval segment, whereas such a marking is absent in males (**Fig.8.6**).

8.6.4 Adult

The adult moth emerging from the pupa is incapable of flight as a result of its domestication for more than four thousand years. It does not feed during its short adult life. The body of the moth like that of the larva is composed of three distinct segment common with other moths and butterflies are situated on either side of the head. The ocelli are absent. The antennae are bipectinate. The thorax consists of three segments namely pro-, meso- and meta-thorax. Mesothorax is the largest and is pentagonal. There are three pairs of thoracic legs, one pair on each of the three thoracic segments. Each of the thoracic legs is composed of five segments. The mesothorax bears two pairs of wings, the front pair overlapping the hind pair when the moth is in the resting position. In the male eight abdominal segments are visible; in the female present laterally on either side of the body. Morphologically the female and male can be easily distinguished in the adult stage. Compared to the male moth, the female has smaller antennae and a fatter body and abdomen. In order to prepare hybrid eggs, the male moth has a pair of knob-like projections at the caudal end that include sensory hairs dividing the sexes. **(Fig.8.7, Fig.8.8)**

Fig. 8.7: Adult of silkworm

Fig. 8.8: Female and male moths

References

Chapter morphology and life cycle of silkworm species and their host plants. Slide Share. https://www.slideshare.net/Nayananayu2/2-a-chapter-morphology-and-life-cycle-of-silkworms-species-and-their-host-plants

Life cycle and morphology of Bombyx Mori. (n.d.-a). https://hbmahesh.weebly.com/uploads/3/4/2/2/3422804/life_cycle_and_morphology_of_bombyx_mori._word.pdf

Outcomes Assessment

Part A

Answer the following questions true or false. [1×5=5 Marks]

1. Herold's gland is found in females. True/False
2. In 48-72 hours spinning gets completed. True/False
3. Larval stage is easier for determining sex. True/False
4. Silk glands are tiny in silkworms. True/False
5. Pupal period of silkworm may vary from 8-14 days. True/False

Part B

Answer the following questions. [2×5=10 Marks]

1. Life cycle of silkworm.
2. Moulting in silkworm.
3. Characteristics of eggs of mulberry silkworm.
4. Characteristics of head of mulberry silkworm.
5. Characteristics of larvae of mulberry silkworm.

Part C

Write a brief note on each of the following. [5×2=10 Marks]

1. Describe briefly about the life cycle of silkworm with a neat diagram.
2. Describe about the morphology of silkworm at various stages.

Part D

Select the correct answer from the four choices given under each question.

[1×4=4 Marks]

1. In how much time spinning of cocoons get completed?

   a. 12 hours
   b. 24 hours
   c. 24-48 hours
   d. 48-72 hours

2. Spinneret is found at which stage?

   a. Egg
   b. Larvae
   c. Pupae
   d. Adult

3. Duration of instar and moulting period in multivoltine races?

   a. 5 days
   b. 12-10 days
   c. 22-23 days
   d. 26 days

4. Duration of instar and moulting period in uni and bivoltine races?

   a. 21-20 days
   b. 35 days
   c. 24-26 days
   d. 15-19 days

Answers Key: 1. **d, 2. b,** 3. **c,** 4. **c.**

9

Study of Silk Glands and Digestive System of *Bombyx mori* L.

9.1 Structure of Silk Gland

The silk of silkworms is secreted by a pair of labial glands, known as silk glands. The silk glands lie ventral to the alimentary canal. In full grown larvae, these occupy most of the body cavity. The silk glands are tubular in shape with different diameters in different regions. Each gland has 3 distinct regions (**Fig.9.1**).

(1) **Posterior Region:** Blunt, highly folded tubular posterior regions of both glands remain attached to tracheal bushes of silkworm. This part secretes fibroin as fibrinogen which converted to fibroin upon extrusion.

(2) **Middle Region:** Most prominent and widest part of silk gland. It remains folded in a W-shaped structure and thus has 3 limbs posterior, middle and anterior limbs. The posterior arm secretes sericin-I. It gets surrounded by sericin-II secreted from the middle limb. This sericin again gets surrounded by sericin- III secreted from the anterior limb. The middle region of silk gland also acts as the reservoir of fibroin where the later gets mature during the storage period.

(3) **Anterior Region:** The thin anterior region of silk gland has no secretory role and only transports the assembled silk to the spinneret.

Spinneret

It is a projection of the median part of the labium, which draws the silk out in the form of fine filament. The secreted silk comes out as a thread or filament as it passes through silk press which resembles a typical salivary pump. The two filaments coming out of two sides are called brins. The sericin (gum) layer of the two brins then bind together into a single filament or bave.

Histologically the entire gland has 3 layers

(1) The outer tunica propia with uniform thickness;

(2) The middle glandular layer with gland cells which increase in size during later instar stages of larval development and

(3) The inner tunica intima: It has varying thickness.

In the anterior region of the gland, this layer is very thick and is shed at each moult. In other regions of silk gland, it is thin and not shed at each moult.

Filippi's gland or Lyonnets's gland

In the head region of the larvae, a pair of glands is situated which open into the anterior part of silk gland near its opening into the spinneret. It is thought that these glands contribute some waxy materials to the silk thread or lubricate the passage of silk while coming out.

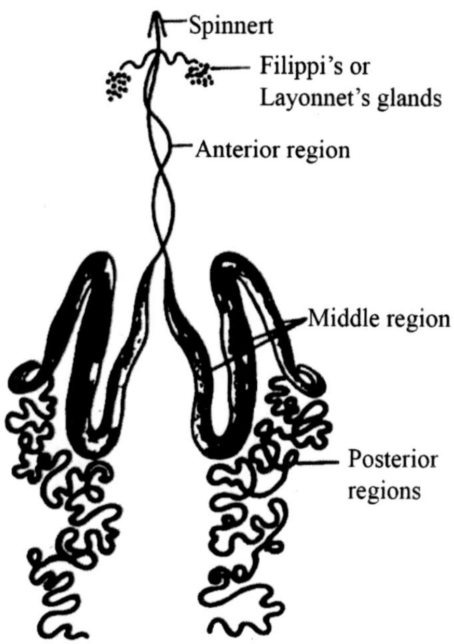

Fig. 9.1: Silk glands of silk worm larva

9.2 Digestive System of Silkworm Larva

In the silkworm larva the digestive system is more or less a straight tube from the mouth to the anus divided into three main parts: the fore-gut or stomodeum, mid-gut or mesenteron, and hind-gut or proctodeum. The oral aperture opens into the mouth cavity which is followed by a narrow pharynx and oesophagus.

The oesophagus is narrow at the anterior end and gradually widens towards the posterior end. There is a cardiac or stomodeal valve at the end of the fore-gut that retains the chewed mulberry leaf bits in the oesophagus for some time and also prevents the regurgitation (backward flow) of food from mid-gut to fore-gut. The mid-gut is a long, wide cylindrical tube narrow at the posterior end. Digestion and assimilation of food, take place mainly in the mid-gut. The digestive fluid is secreted principally from the goblet cells at the mid-gut epithelium and the cylindrical cells absorb the digested food. The fore and hind guts have a chitinous lining, but an inner layer of peritrophic membrane in the mid-gut generally protects the mid-gut epithelium from mechanical damage due to food particles. The hind-gut consists of the small intestine, colon and rectum, and a pylorus valve near the anterior end of the small intestine which guards and regulates the passage of digested food from the mid-gut to the hind-gut. The hind gut is a passage for the absorption of a large portion of food moisture and elimination of digested food. In the anterio-posterior direction, the rectum has six muscles for pressing the excrements. The faecal matter is pressed in the rectum and expelled from the anus as faecal pellets bearing hexagonal marks (**Fig.9.2**).

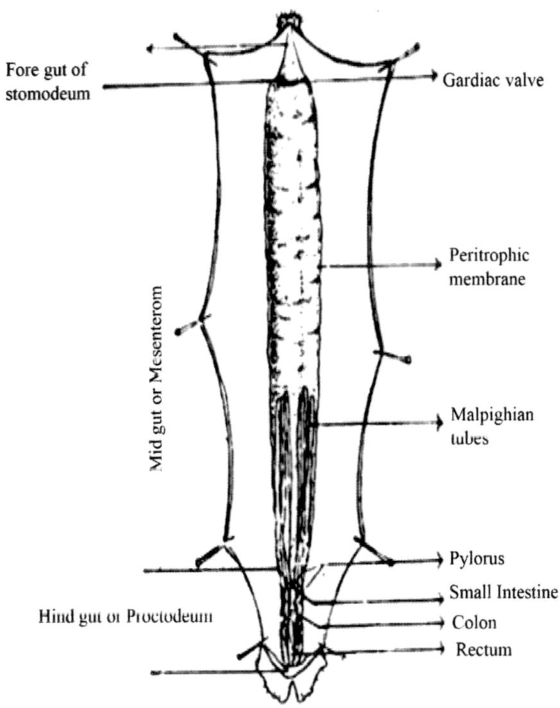

Fig. 9.2: Digestive system of silk worm larva

References

Sericulture Manual II, FAO, Rome, 1987.
The Principles of Insect Physiology by V B Wiggelesworth, 1972.
The silkworm – A Laboratory Tool by Y Tazima, 1979.

Outcomes Assessment

Part A

Answer the following questions true or false. [1×5=5 Marks]

1. Silk is secreted by one labial gland. True/False
2. The entire gland has 3 layers. True/False
3. Middle region of the gland is folded as Z shaped. True/False
4. Peritrophic membrane protects from mechanical damage. True/False
5. The faecal matter bears hexagonal shape. True/False

Part B

Answer the following questions. [2×4=8Marks]

1. Describe about the 3 regions of silk gland.
2. What is Lyonnets's gland?
3. Describe the Mechanism of silk production.
4. Describe about the mechanism of digestion in silkworm.

Part C

Write a brief note on each of the following. [5×2=10 Marks]

1. Describe briefly about the mechanism and structures present in silk gland with a neat labelled diagram.
2. Describe briefly about the mechanism and structures present in digestive system with a neat labelled diagram.

Part D

Select the correct answer from the four choices given under each question. [1×4=4 Marks]

1. Faecal matter bears what shaped markings on it?

   a. Triangle b. Hexagonal
   c. Circulator d. None of the above

2. Silk secreted by how many labial glands?

   a. 1 b. 2

   c. 3 d. 4

3. Middle region of silk gland is folded in which shaped?

   a. W b. Z

   c. H d. Y

4. Which layer of silk gland have uniform thickness?

   a. Outer tunica propia b. Middle glandular

   c. Inner tunica propia d. None of the above

Answer Key: 1. **b,** 2. **b,** 3. **a,** 4. **a.**

10

Disinfectants: Rearing Bed and General Disinfectants

10.1 Disinfection

Disinfection is the destruction of disease carrying germs. Proper disinfection is one of the key factors in determining the success of sericulture and also essential in prevention and control of silkworm disease. The disease caused by virus, bacteria, fungi and protozoa are known to occur in almost all the silkworm rearing areas of the world. The disease, together with the damage caused by the pest, accounts for 20 to 25% of crop loss.

10.2 Qualities of an Ideal Disinfectant

- It should be highly effective against wide variety of microorganism and non-toxic for animals and plants.
- It should be non-injurious and non-staining to materials like fabrics, furniture or metal wares and non-offensive to odours and taste.
- It should be specific for specific microorganisms.
- It should have good surface tension. That is good wetting and penetrating properties.
- It should be stable in storage.
- It should be readily available in market and non-expensive.
- It should be easily applied under household or other practical conditions of use.

10.3 Methods of Disinfection

There are 2 types of disinfection methods:

10.3.1 Physical Methods

The physical methods of disinfection include simple and effective physical treatment for prevention and control of diseases.

It includes the following methods:

10.3.1.1 Burning and Burrying

This is the simplest method of disinfection, where the diseased specimen that is eggs, larva, pupa, and adult or contaminated particles are burnt or burried deep underneath soil.

10.3.1.2 Exposure under Sunlight

Direct disinfection by exploiting the natural sunlight can be considered as supplementary measure for pathogen disinfection. The germicidal effect of sunlight derived from UV radiation kills the pathogen. Exposure of contaminated rearing appliances to sunlight results in lowering of the disease.

10.3.1.3 Sterilization with Steam

Steaming is another effective and economic method of disinfection for equipment. Steam under high pressure denatures the protein of the pathogen, thereby killing the pathogen.

10.3.1.4 Sterilization by Boiling

In this method, the worst equipment is kept submerged in boiling water for duration of 30 minutes and then dried.

10.3.2 Chemical Method

The chemical affects the metabolism of pathogen through oxidation or reduction of tissue, or by denaturation and precipitation of proteins of the causal organism. It is the most effective method for elimination of germs and is conducted either by spraying or by fumigation.

The effectiveness of chemical disinfection depends upon 3 factors, i.e., concentration, duration and temperature. The chemical disinfection may be broadly classified into room disinfectant and bed disinfectant.

10.3.2.1 Room Disinfectant

The disinfection of rearing house is done 8 to 10 days before starting rearing operation. Some of the chemicals commonly used as room disinfectants are:

- 5% bleaching powder solution

- Mixture of 2% formalin + 0.5% slaked lime
- Sodium hypochloride
- Formalin/Paraformaldehyde
- Chlorine dioxide, e.g., Sanitech

10.3.2.2 Bed Disinfectant

Commonly used bed disinfectants are:

- Labex (composition: 97% Lime and 3% Bleaching powder)
- Formalin
- Dithane M 45
- RKO (Resham Keet Ousadh), Composition: (1% Captan + 1% Paraformaldehyde + 2% Benzoic acid + 96% Slaked lime powder)
- Vijeta/Resham Jyoti @ 5-6 kg/100 dfls (Disease Free Layings)

10.4 Rearing Bed Disinfection

For managing silkworm illnesses in a silkworm colony, it is essential to observe hygiene practices. Those are aimed at restricting ailment unfold within the silkworm colony and prevention of secondary infection. It involves utility of bed disinfectants, to save you the spread of contamination in the colony of silkworms. To keep away from secondary contamination and spread of disease in the colony, silkworm frame and rearing seat disinfectants are to be carried out. Periodic software of the mattress disinfectants as consistent with the recommended time table significantly prevents the spread of illnesses throughout rearing. Software of slaked lime powder at the start of each moult facilitates in soaking up moisture and regulates mattress humidity. Slaked lime being alkaline is a powerful viricide and surprisingly useful in handling the viral infections inside the silkworm colony. Software of mattress disinfectants after every moult is best disorder control exercise. Widespread silkworm body and rearing seat disinfectants are antifungal, antibacterial, antiviral and anti-microsporidian in nature. Application of slaked lime dirt in combination with bleaching powder in and around rearing house and premises improves hygiene in the rearing surroundings.

Disinfectants for silkworm mattresses have long been used to prevent diseases from developing while a silkworm harvest is growing. Several capability mattress disinfectants were developed in India for sickness management all through silkworm rearing. Many popular bed disinfectants are within the subject for controlling numerous illnesses of silkworms. To prevent secondary contamination, prevalence and spread of illnesses in silkworm colony, silkworm body and rearing seat disinfectants are dusted over the silkworm body and rearing seat following a time table. Slaked lime powder is dusted on every occasion the larvae settle for moult. Other commercially available mattress disinfectants are dusted whilst the larvae come out of moult and mulberry is fed after half an hour of dusting the mattress disinfectant. The disinfectants are dusted at 3-5 g/sq. ft. bed region. It is important to understand that mattress disinfectants only stop the transmission of diseases; they do not cure silkworms of ailments. It must then be employed as a safety precaution by evenly dusting the powder in accordance with the suggested quantity and schedule.

To stop all diseases from developing, the mattress disinfectant of choice should be effective against all silkworm pathogens. Even when using mattress disinfectant formulas, certain precautions are necessary. It must no longer be sprinkled on the silkworms during or after their moult or on the edible mulberry leaves that are available in mattresses. Even though the mouth, nose, ears, and eyes should be shielded from dust, arms should be thoroughly cleaned after dusting.

References

Disinfectioninsilkwormrearing.(n.d.-b).-Disinfectioninsilkwormrearing200306050503034444.pdf

Outcomes Assessment

Part A

Answer the following questions true or false. [1×3=3 Marks]

1. Slaked lime is a powerful bactericide. [True/False]
2. Ideal disinfectant should kill wide range of microbes. [True/False]
3. Chlorine dioxide is used as bed disinfectant. [True/False]

Part B

Answer the following questions. [2×3=6 Marks]

1. What are the ideal characteristics of disinfectants.
2. Physical methods of disinfection.
3. Chemical methods of disinfection.

Part C

Write a brief note on each of the following. [5×2=10 Marks]

1. What are the methods of disinfection.
2. Describe about rearing bed disinfection.

Part D

Select the correct answer from the four choices given under each question. [1×4=4 Marks]

1. What percentage of bleaching powder is present in labex?

    a) 5 b) 8

    c) 97 d) 3

2. Diseases account for what percentage of damage?

    a) 50 b) 10

    c) 20 d) 70

3. Commonly used bed disinfectant is

    a) Labex b) Formalin

    c) Vijetha d) All of the above

4. Economic method of disinfection for equipment is

    a) Steaming b) Burning

    c) Drying d) All of the above

Answer Key: 1. **d**, 2. **c**, 3. **d**, 4. **a**.

11

Grainage Techniques

11.1 Grainage

Grainages are establishments where disease free and quality seeds are produced on scientific lines.

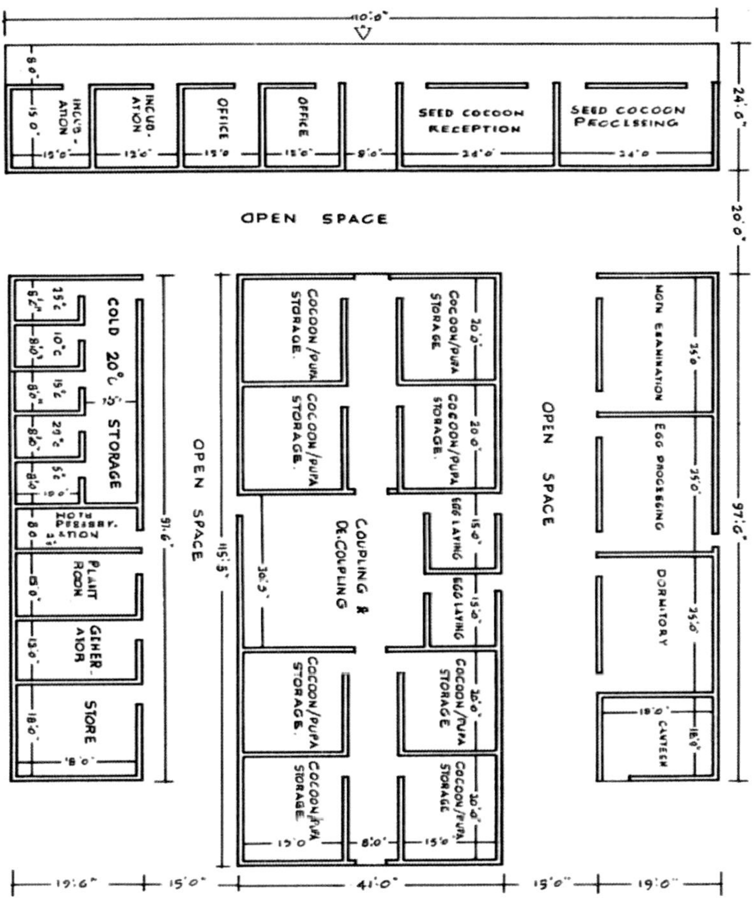

Fig. 11.1: Plan of a grainage building

11.2 Grainage Activities

• Procurement of seed cocoons

Within the cocoon market, the government can buy the necessary number of multivoltine cocoons from the seed cocoon market. The purpose of a grainage is to produce attractive seeds. This method involves unique procedures, such as the manufacture of silkworm seeds for the market. Half of the bivoltine cocoons should also be obtained from bivoltine cocoon marketplaces to guide hybrid seeds.

• The seed cocoons are packed

Loosely in perforated packing containers or bamboo baskets in small portions and are transported all through cooler hours of the day. Initially the seed cocoons arriving on the grainages are subjected to inflexible selection. In selection only sound and uniform cocoons conforming to the characteristics of the race are selected and defective cocoons and so forth., are rejected. Strengthen detection of pebrine disease, if any before the commencement of operation of each batch enables in keeping off excellent loss to the grainages. This is facilitated by means of investigations at 3 stages:

1. Pupal test
2. Forced eclosion check
3. First day moth examination.

The pupa is reduced ventrally just below the wing bud by way of a scissor by means of keeping the pupa between thumb and fore finger in left hand. After cutting the pupa is pressed gently. The midgut oozes out as a brown frame from the reduced portion. This midgut is collected and crushed with few drops of potassium hydroxide in a moth crushing set. The fluid is taken at the slide and study below the microscope with 600 X magnification.

• Protection of seed cocoons

The cocoons are kept in single layers in well-ventilated rooms with herbal lighting, as well as areas with natural light and darkness. Exhaust fans in cocoon protection and emerging rooms are crucial for removing bad gases and dust. Unique races will be protected in distinct chambers. In the event of business hybrid seed propagation, the sexes should be separated before selfing occurs within the same parental race. This sexing can take place at the larval, pupal or moth level. Moths of the same races are made to emerge on the identical day, so that male and female moths are quite simply available for hybridization.

That is referred to as synchronization. In any other scenario, the emergence of moths in the two races can be regulated by picking cocoons of the same date. To postpone the emergence of early batches during synchronization, cocoons can be refrigerated at 5-10°C. Such refrigeration must be limited to three days for female and seven days for male. The moths will be refrigerated at 5°C for up to ten days in adult males and 2-3 days in females.

11.2.1 Coupling and Decoupling

As quickly as the moths emerge, they begin to get pair. The female moths are unfold in a tray and the male moths of the perfect hybrid element are flippantly disbursed over the females. In about 15 minutes, the male and female moths pair. After the desired period of coupling, the pairs are separated through protecting the female moth and gently sliding the male. This facilitates clean separation without injury to the female.

Oviposition: This is not anything however the procedure of egg laying via the female moth. While eggs are required to be laid on playing cards, the mated females are located on the egg sheets and every moth is enclosed in a cellule. The cellule isolates the eggs laid by using each moth, facilitates male or female moth examination and removal of eggs laid via diseased moth.

In case of preparation of free eggs, a unit range of female moths are allowed to lay eggs on starched paper or fabric, within a timber or plastic body. The number of moths varies between 30 and 200 depending on the batch's fitness and comfort. Moths are allowed to lay eggs for 24 hours in a darkish room with 25°C and 75-80% RH. Then mother moths are examined for the pebrine ailment and eggs free from pebrine disease are qualified for rearing.

- **Mother moth examination**

Individual mother moth examination is done by sample testing.

Surface Disinfection of Eggs: After mother moth examination, the egg sheets or free eggs in a box are dipped in 2% formalin solution.

1. This allows in putting off surface contamination.
2. This facilitates in casting off surface infection.
3. Formalin will increase the adhesive capability of eggs to the egg sheets.

Subsequently, the sheets are washed and dried beneath shade and preserved underneath top-quality temperature (25°C) and RH of 80%.

11.2.2 Artificial Hatching

Multivoltines by no means undergoe diapause, but uni- and bi-voltines go through diapause or hibernation. So, to test the diapause artificial hatching is carried out via either cold remedy or Hydrochlorination cold remedy is not anything however refrigeration relying upon time table i.e., hibernation schedule.

Hydrochlorination is hydrochloric acid treatment:

1. Cold acid treatment: HCl 1.1 sp.gr.60-90 min at 24°C
2. Hot acid treatment: HCl 1.075 sp.gr. 5-6 min at 46.1°C

Then rinse the eggs in running water to remove acid traces and dry under shade.

Reference

Manual on silkworm egg production, Central Silk Board, India 1988.

Outcomes Assessment

Part A

Answer the following questions true or false. [1×5=5 Marks]

1. Crossing occurs in equal parental race. [True/False]
2. Grainages are subjected to flexible selection. [True/False]
3. Refrigeration confines 7 days for females. [True/False]
4. Refrigeration confines 7 days for male. [True/False]
5. Every moth is enclosed within cellule. [True/False]

Part B

Answer the following questions. [2×3= 6 Marks]

1. Describe about the procurement of seed cocoons.
2. Describe about the protection of cocoons.
3. Describe about the mother moth examination.

Part C

Write a brief note on each of the following. [5×2=10 Marks]

1. Describe about the 3 steps of mass examinations made.
2. Briefly describe about artificial hatching.

Part D

Select the correct answer from the four choices given under each question. [1×4= 4 Marks]

1. Protection of seed cocoon is done at a temperature of?

   a. 25°C b. 30°C

   c. 20°C d. None of the above

2. Protection of seed cocoon is done at a relative humidity of?

   a. 90 b. 70

   c. 60 d. 80

3. What is the duration of hot acid treatment?

   a. 30 mins b. 20 mins

   c. 10 mins d. 5 mins

4. Moths are allowed to lay eggs for?

   a. 1 day b. 2 days

   c. 7 days d. 5 days

Answer Key: 1. **a**, 2. **d**, 3. **d**, 4. **a**.

12

Study of Rearing House Plan and Equipment for Shoot Feeding and Shelf Rearing

12.1 Introduction

Mulberry silkworm is a domesticated insect which cannot live without sufficient human care. It is also sensitive to high temperature and relative humidity. Besides good feed, a temperature range of 23°C to 28°C and relative humidity of 75-80% is ideal for silkworm. Though, there are a number of modern gadgets to regulate the temperature and relative humidity, investment on them is uneconomical. Therefore, mulberry sericulture is practiced only in those areas where the climate is congenial and the crops schedule is so regulated to avoid seasons with extreme climate. Besides, the silkworm rearing house is constructed to provide sufficient ventilation and keep the ambiance sufficiently cool. Due to poverty, many farmers rear silkworms within their dwelling house which cannot be disinfected due to the pungent nature of the chemicals used. However, it will be ideal to have a separate rearing house or at least a separate room for rearing purpose. The present unit deals with different kinds of rearing houses required for silkworm rearing.

12.2 Characteristics of Rearing House

A rearing house should essentially provide sufficient bed space for silkworms and working space for the workers attending the rearing operations, good ventilation to replacc ammonia, carbon dioxide and other noxious gases released by the silk worms during respiration, excretion, etc. There should be space for leaf preservation, storing other appliances and chemicals or disinfectants used in rearing.The rearing house should be provided with mesh or nylon net all around to prevent entry of uzifly, the most dreaded pest of silkworm. An ante-chamber is necessary to avoid the entry of uzifly, so that the workers can watch, if the fly enters along with them. Lizards and rats are the common predators and therefore construction of a rat-proof sill around the rearing house is necessary. The rearing house should facilitate making it air tight for fumigation or disinfection to ensure the fume or vapour from the disinfectants remain inside the house till the germs inside are completely eliminated. Most of the sericulture area falls under warmer zones and the rearing house is expected to provide a cooler ambience to the silkworm. Use of any material that keeps the inner environment cool and prevents radiation from the sun is preferred. Mud wall and thatch roof are the most ideal that can cool the inner ambience to the maximum, though, its life is shorter and requires repair of the roof once in every 3-4 years. Hollow cement bricks, burnt earthen bricks and stone slabs are also used very commonly.

North or south facing of the rearing house is preferred as it provides good aeration since the wind blows either from the north-east or from the south-west. This also reduces the chance of direct sunlight falling on the rearing bed either in the morning or in the afternoon.

12.3 Selection of Site

As we use pungent disinfectants, which may be hazardous particularly to those suffering from respiratory problems, it is ideal to have the rearing houses away from the thickly populated areas. It is also advisable to keep them off the places of livestock as the chemicals used may harm them. It is also necessary to avoid damp areas which facilitate easy multiplication of disease causing germs and spread of diseases. Areas closer to tobacco barrens, lime or brick kilns, and industrial establishments emanating dust, smoke and other noxious gaseous are unfit for rearing activities. Keeping them close to gardens or fields where excessive pesticides are sprayed is also not advisable. Shaded areas under large trees are good for locating rearing houses as they provide cool environment and fresh and clean air in the areas where the temperature and humidity is high. It is ideal to have the rearing house closer to the mulberry garden as it will be convenient to feed fresh leaves with very little moisture loss that occurs during transportation.

12.4 Accommodation for Different Activities in Rearing

The rearing activities involve Chawki rearing, i.e. rearing of young worms usually of the first two instars; late age rearing, involving the rearing of third to 5th instar larvae and mounting for spinning. Small farmers use the same rearing house for all the activities which together lasts for 30-35 days including cleaning and disinfecting. Larger farmers use separate rearing houses for rearing young (Chawki) worms and separate mounting hall.

Chawki Rearing

Shelf Rearing

Tray Rearing

Fig. 12.1: Different types of rearing

As young larvae (up to 2nd moult) require to be fed with tender chopped leaves and the bed humidity maintained high (80-85%) and cared much, they are usually reared in wooden or plastic trays with the worms wrapped in paraffin paper with one sheet on the floor of the tray and the other above the silkworms fed with the chopped leaves and the trays placed one above the other in tiers. Around 30-40 square feet of bed area is required for various breeds of silkworms up to the end of 2nd instar. The space requirement is double if reared up to the end of 3rd instar. The success of the cocoon crop depends up on the care taken during the Chawki stage. For maintaining the temperature and humidity, small and compact room is required to rear young age silkworms. Hence, the concept of community Chawki rearing under expert supervision and special care is being promoted. A viable Chawki rearing centre shall rear from 2500 to 5000 layings per batch which runs for 8 days.

Late age rearing is done in two methods, *viz.*, tray (leaf) feeding associated with leaf plucking and shoot feeding associated with shoot harvest. In case of leaf feeding, around 38 m^2 of bed area is required for rearing 100 layings. Since the rearing is done in rearing trays placed on rearing stands in tiers each with 10-12, it requires a rearing house with a smaller floor area (around 20 m^2). However, it requires a higher manpower for feeding operation. Around 70 m^2 bed area is required for rearing 100 layings in case of shoot feeding. This is organized by erecting rearing shelves in 4 or 5 tiers. A platform of 5' × 35' will accommodate 25 layings. A rearing house with floor area of 400 sq. ft. will be enough to rear 100 layings. Rearing rooms/halls are constructed to accommodate the rearing stands or rearing racks lengthwise, one set on either side of the room with a reasonable working space.

Fully mature larvae stop feeding, empty their gut and become translucent. They crawl (for support) for spinning cocoon around them. At this stage, the worms are transferred to the mountages. The traditional bamboo-made mountages are called 'chandrike'. They can also spin cocoons on straw or dry twigs. There are a number of improved mountages like plastic collapsible mountages and rotary cardboard or plastic mountages where the wastage of silk is less. Spinning and pupation lasts for 4 days after which the cocoons are harvested from the mountages and taken to market. Before spinning, the silkworms urinate causing a high humidity and pungent smell. Therefore, the spinning worms require a dry climate with free flow of air to ensure formation of stiff cocoons without urination spots. Therefore, the mounting halls are almost near open sheds with wide windows and with wire mesh covering all around.

The environmental condition required for leaf preservation is different from the silkworm rearing and there is chance of mixing of leaf or shoot with the silkworm litter if leaf is preserved in the same room where rearing is being conducted. Hence, separate room is required for preserving the leaf/shoot. It is suggested to have a leaf preservation room with each rearing house as the leaf has to be preserved without moisture loss and free from contamination from the rearing bed refuse. Also, it is advised to have a small disinfection tank to dip the rearing trays and mountages in the disinfectant solution so that their disinfection shall be complete.

References

A Guide for Construction of Silkworm Rearing House - Published by Member Secretary, Central Silk Board, Bangalore, 1999.

Ganga, G. 2003. Comprehensive Sericulture - Vol. II, Silkworm Rearing and Silk Reeling, Published by Oxford and IBH

Rajan, R.K. and Himantharaj, M.T. 2005. A Textbook on Silkworm Rearing, Published by Central Silk Board, Bangalore.

Seri-business Manual - A Users' Guide (Farm sector) - Published by Member Secretary, Central Silk Board, Bangalore, 2003.

Outcomes Assessment

Part A

Answer the following questions true or false. [1×4=4 Marks]

1. Up to 3rd moult the young larvae has to be fed with tender chopped leaves. True/False
2. Ideal temperature range for silkworm is 10-15°C. True/False
3. Chawki rearing is the rearing of young larvae. True/False
4. Rearing house activities last for 35-40 days. True/False

Part B

Answer the following questions. [2×3=6 Marks]

1. How to select site for rearing?
2. What are the characteristics of rearing house?
3. What are the different activities done in rearing house?

Part C

Write a brief note on each of the following. [5×2=10 Marks]

1. What are the different activities taken under rearing? Describe briefly.
2. Describe briefly about the characteristics of an ideal rearing house?

Part D

Select the correct answer from the four choices given under each question. [1×4=4 Marks]

1. A platform of 5' × 35' will be enough to rear?

   a. 50 layings b. 300 laying

   c. 100 layings d. 200 layings

2. Spinning and pupation lasts for how many days?

   a. 6 days b. 10 days

   c. 14 days d. 4 days

3. Viable Chawki rearing centre shall rear how many layings per batch?

   a. 10000 b. 2500

   c. 1000 d. 5000

4. Chawki rearing runs for

   a. 14 days b. 4 days

   c. 8 days d. 1 day

Answer Key: 1. **c**, 2. **d**, 3. **b**, 4. **c**.

13

Methods of Incubation of Silkworm Eggs and Brushing

13.1 Incubation

The silkworm eggs are to be incubated properly to avoid irregular hatching. An ideal incubation is to maintain a temperature of 25°C and 75-80% humidity continuously. In most of the places incubators are not available. In such cases, the eggs are kept spread in wooden trays provided with paraffin paper for base and the cover and foam pads kept wet all around inside the trays to maintain 70-80% humidity. In each chawki tray of 90 cm × 120 cm about 4 sheets of eggs can be kept for incubation. The room temperature is maintained at 25°C. On 9th day the egg sheets are rearranged. About 25 dfls (disease free layings) are kept in a chawki tray of 90 cm × 120 cm to enable brushing and proper spacing for young worms later. The egg sheets are covered with a sheet of black paper or paraffin paper. On the due date of hatching and when a few larvae have started hatching, black paper and the paraffin paper are removed and the eggs are exposed to dim light for about an hour. This facilitates the uniform hatching of larvae.

13.1.1 Methods of Incubation

There are two methods, i.e., constant temperature incubation and raised temperature incubation.

1. Constant Temperature Incubation

In the incubation room optimum humidity should be 80-85%. Optimum temperature for the incubation of non-hibernating eggs and eggs after acid treatment for immediate hatching should be 24-25°C right from the beginning.

2. Raised Temperature Incubation

It is used mostly for incubating hibernating eggs. The eggs after release from cold storage, are preserved at 10-15°C for three days, at 18-20°C for two days, at 23-24°C for four days and then at 25-26°C till hatching.

13.2 Brushing

The transfer of black and bristly newly hatched larvae called the "ants" from the egg shells to the rearing bed is referred to as brushing. Since hatching is a photoperiodic response, hatching commences in the early morning and continues till noon. Hence, the best time for brushing is 10.00 a.m., i.e., the peak time of hatching. The unhatched eggs are transferred to the egg box or black box and brushed the next day. The larvae brushed for two successive days are reared together and the larvae hatched thereafter have to be either discarded or reared separately.

(a) Brushing from loose eggs in egg boxes

When 75% of the eggs reach the blue egg stage, the cover of the egg box or black box is removed and a thin muslin cloth or minutely perforated paper sprinkled with chopped mulberry leaves is placed just above the eggs on the inner side of the paper or cloth. When the newly hatched larvae crawl onto the cloth, the larvae along with the chopped leaves are transferred to the rearing bed containing fresh leaves.

(b) Brushing from egg cards

(i) **Feather method:** The egg card with newly hatched larvae is held vertically above the rearing bed and the larvae can be gently brushed from the egg shells to fall on the rearing bed. This method may injure the larvae by dashing them against the cut edges of the egg shell or chorion.

(ii) **Husk method:** Powdered husk is sprinkled over the larvae on the egg card and chopped mulberry leaves are sprinkled at the centre of the husk. The larvae crawl up to the centre of the husk for the leaves. After 30 minutes, the leaves with larvae can be brushed to the rearing bed. This method minimizes the injury of the feather method but the disadvantage is that a little quantity of husk invariably reaches the rearing bed.

References

Sericulture. (n.d.-i). https://bieap.apcfss.in/Uploads/Materials/II_SERICULTURE.pdf

The Insects: Beneficial and Harmful Aspects," Proceedings of the Entomological Society of Washington 111(2), 544-545, (1 April 2009).

Rajan, R.K. and Himantharaj, M.T. 2005. A Textbook on Silkworm Rearing,Published by Central Silk Board, Bangalore.

Outcomes Assessment

Part A

Answer the following questions true or false. [1×5=5 Marks]

1. On the 4th day egg sheets are rearranged. True/False
2. In constant temperature incubation hatching should done at a temperature of 10°C. True/False
3. Powdered husk is sprinkled in feather method. True/False
4. Newly hatched larvae are called ants. True/False
5. Chawki tray keeps 4 sheets of egg trays. True/False

Part B

Answer the following questions. [2×3=6 Marks]

1. Describe the methods of incubation.
2. Describe the brushing from egg cards and its types.
3. Describe the brushing of loose eggs from egg box

Part C

Write a brief note on each of the following. [5×2= 10 Marks]

1. Describe briefly about brushing and methods of brushing.
2. Describe briefly about incubation procedure and methods of incubation.

Part D

Select the correct answer from the four choices given under each question. [1×3=3 Marks]

1. Newly hatched larvae are known as
   - a. Caterpillar
   - b. Ants
   - c. Grub
   - d. Maggot
2. What is the temperature range kept at constant temperature incubation?
   - a. 28°C
   - b. 25°C
   - c. 20°C
   - d. 30°C
3. What is the best time for brushing?
   - a. 12.00 a.m.
   - b. 3.00 p.m.
   - c. 10.00 a.m.
   - d. 9.00 p.m.

Answer Key: 1. **b**, 2. **b**, 3. **c**.

14

Identification of Silkworm: Settling for Moult, at Moult and out of Moult

14.1 Larval Life

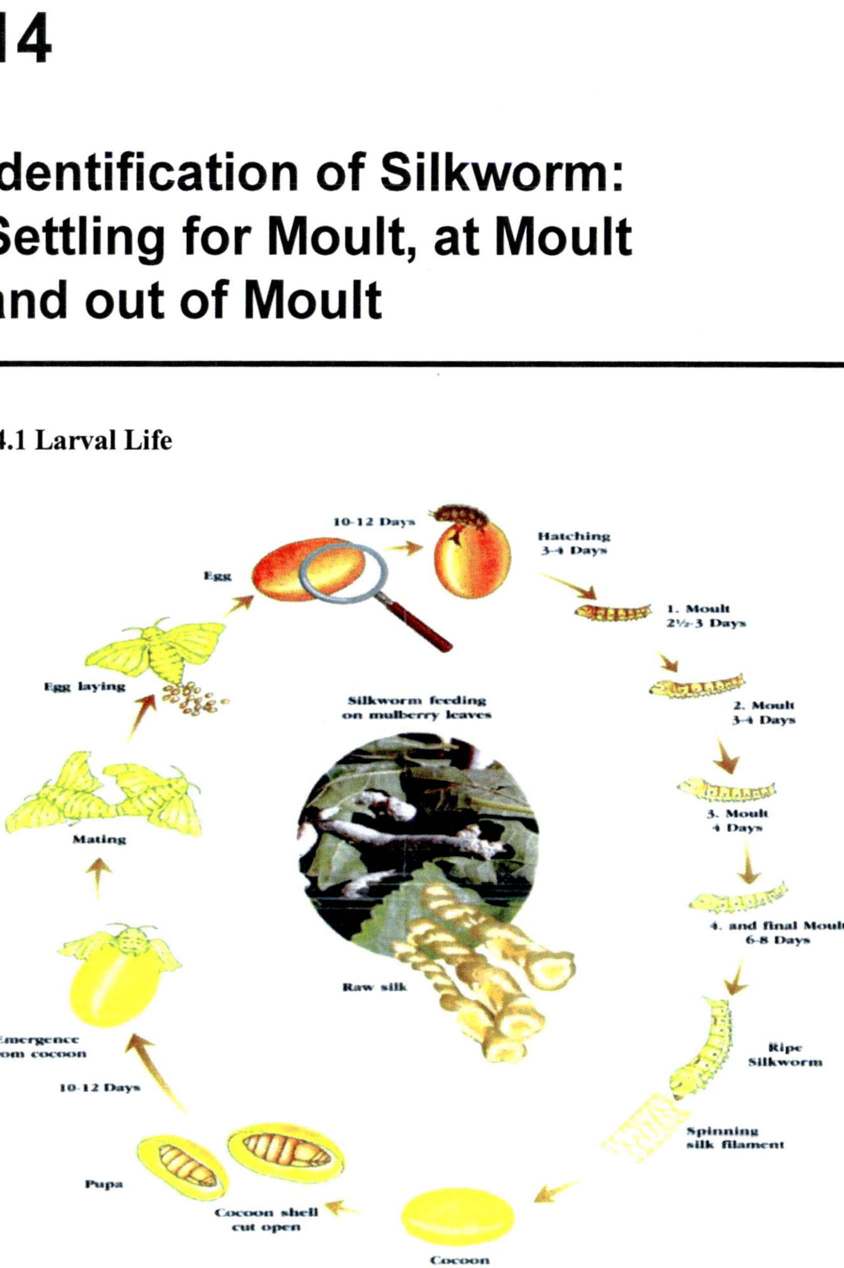

Fig. 14.1: Life cycle of silkworms (Bombyx mori)

In sericulture it is the larval life that is of direct importance to the rearer since he has to take care of the worms very carefully. The larval life may last from

20-24 days in the case of the multivoltine species in tropical areas or 24-28 days in the case of uni- and bi-voltine races in temperate areas, being shorter under warmer summer and autumn conditions, and somewhat longer under cooler spring conditions. During the larval life the worm moults or casts off its skin four times to be able so as to grow. There are also varieties of worms which moult three and five times but they are not of importance economically. In view of the four intervening moults, the larval life is divided into five distinct stages or instars (**Fig.14.2**) which are referred to popularly as five different ages. The first three instars are referred to as young ages and the fourth and fifth instars as late ages. The duration of the different instars and moulting periods (**Table 14.1**) for the different races is as follows:

Table 14.1: Duration of different instars and moulting periods

Duration of different instars and moulting periods					
		Multivoltine races		**Uni and bivoltine races**	
	Duration		**Temperature and humidity conditions**	**Duration**	**Temperature and humidity conditions**
I	instar	3 days	27°C 80-85% R.H.	3 days	27"C and 85% R.H.
I	moult	20 hours		20 hours	
II	instar	2 days		2 days	
II	moult	20 hours		20 hours	
III	instar	3 days		3 days	25°C and 30% R.H.
III	moult	1 day		1 day	
IV	instar	4 days	25-26°C 70-80% R.H.	5 days	22-24°Cand 75% R.H.
IV	moult	1 day		1 day	
V	instar	6-7 days		9-10 days	20-23°Cand 70%
Total Duration		22-23 days		26-27 days	

14.2 Moulting

Each larval instar can be broadly divided into two phases i.e., the feeding phase and the moulting phase. After feeding voraciously and having attained full growth for the particular instar the worm loses its appetite and the larva prepares to moult and cast off its old skin. Prior to each moulting, the larva stops feeding and rests with its head held up. The lustrous body skin gradually becomes translucent, loose and wrinkled and the worm becomes dull in appearance and wanders about in search of a resting place. It emits a silky substance with which it fixes itself to dry leaves. The moulting periods last for 20 hours being shortest in the second moult, followed by the first, third and fourth. This resting for moulting is often referred to as going to sleep and the coming out of the worm from moult as waking up. During the moulting period the worms should not be disturbed so that the process of moulting is not interfered with and uniform moulting is ensured.

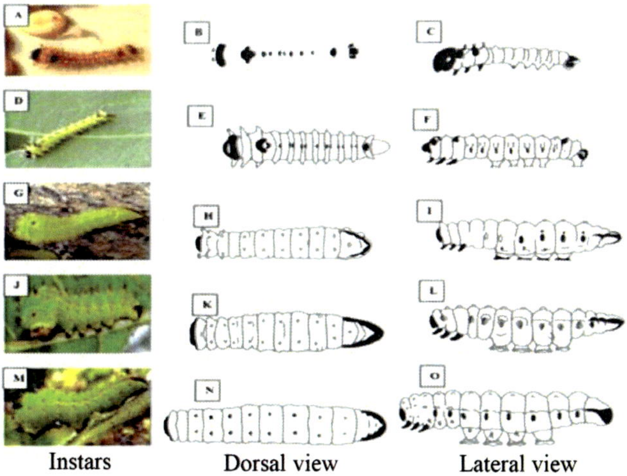

Fig. 14.2: Five instars of silkworm

14.3 Abdomen of Larvae

The abdomen is comprised of eleven body segments although only nine can be distinguished as the last three are fused together to form the apparent ninth segment, the anal plate and the caudal legs. The third to sixth and the last abdominal segments each bear a pair of abdominal legs which are fleshy, unjointed muscular protuberances. At the extremity they form a sort of disc with a series of hooks inwardly curved and arranged in a semicircle. On the dorsal side of the eighth abdominal segment, the larva carries the caudal horn. The abdominal segments carry the sexual markings which develop distinctly in the fourth and fifth instars in the eighth and ninth segments on the ventral side. In the female the sexual markings appear as a pair of milky white spots in each of the eighth and ninth segments. The pair of spots on the eight segments is known as Ishiwata's fore glands and the pair on the ninth segment is referred as Ishiwata's hind glands. In the male a small milky white body (Herold's gland) appears at the centre of the ventral side between the eighth and ninth segments. There are nine pairs of spiracles placed laterally on either side of the body. They are found on the first thoracic segment and the first eight abdominal segments. These are the breathing or respiration pores. The larval skin or integument consists of the cuticle and the hypodermis. The cuticle is made of chitin as well as protein and is covered with a thin layer of wax. Nodules are found all over the body surface of the silkworm larva. The distribution of these nodules differs according to the variety of silkworms, being less in number in varieties having lustrous body surfaces, such as the Chinese bivoltine and the multivoltine varieties.

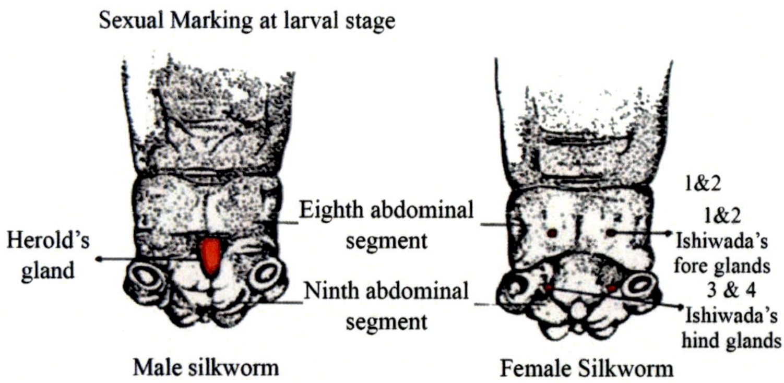

Fig. 14.3: Sexual marking at larval stage

The nodules are found in larger numbers over parts of the body where normally the larval markings are found. The larval markings in silkworms are caused by skin pigment.

Male larva

(i) A milky white body appears at the centre of the ventral side between 8th and 9th abdominal segments. This is known as **Herold's gland.**

(ii) The cocoon produced from the larvae is light in weight.

Female larva

(1) A pair of milky white spots are present in each of the 8th and 9th ventral abdominal segments. These two pairs of spots are known as **Ishiwata's glands.**

(ii) The cocoon produced from the larvae is heavy in weight.

Filippi's gland or Lyonnets's gland

In the head region of the larvae, a pair of glands is situated which open into the anterior part of silk gland near its opening into the spinneret. It is thought that these glands contribute some waxy materials to the silk thread or lubricate the passage of silk while coming out.

References

Life cycle and morphology of Bombyx Mori. (n.d.-a). https://hbmahesh.weebly.com/uploads/3/4/2/2/3422804/life_cycle_and_morphology_of_bombyx_mori._word.pdf

David, B.V. and Ananthakrishnan, T.N. 2004. General and applied entomology (2nd ed.). New Delhi, India: Tata McGraw-Hill Publishing Company Limited, p. 757-769.

David, B.V. and Ramamurthy, V.V. 2011. Elements of economic entomology. Chennai, India: Namrutha Publications, pp.386.

Ganga, G. and Chetty, J.S. 1997. An introduction to Sericulture. (2nd ed.). New Delhi, India: Oxford and IBH Publishing Company Private Limited, pp.302.

Outcomes Assessment

Part A

Answer the following questions true or false. [1×4=4 Marks]

1. Ishiwata glands are found in male silkworms. True/False
2. Herolds glands are found in female silkworm. True/False
3. Duration of instar and moulting period is high in multivoltine species. True/False
4. Filippi's gland is found in the posterior part of the larva. True/False

Part B

Answer the following questions. [2×3=6 Marks]

1. Describe about the larval life of silkworm.
2. Describe about the stages of moulting in silkworm.
3. Explain the sexual dimorphism found in larval silkworm.

Part C

Write a brief note on each of the following. [5×2=10 Marks]

1. Briefly describe about the stages and duration of moulting and instars in larvae of silkworm .
2. Describe about the abdomen Of larva and differences in male and female moth larvae.

Part D

Select the correct answer from the four choices given under each question. [1×3=3Marks]

1. Total duration of instar and moulting in multivoltine race is:

   a. 15 days b. 17 days

   c. 22 days d. 30 days

2. Total duration of instar and moulting in uni and bivoltine races is:
   a. 20 days
   b. 30 days
   c. 18 days
   d. 26 days
3. Herold's gland is found in between which segments of abdomen?
   a. 7th and 8th
   b. 6th and 7th
   c. 9th and 10th
   d. 8th and 9th

Answer Key: 1. **c**, 2. **d**, 3. **d**.

15

Bed Cleaning, Feeding and Spacing

15.1 Silkworm Bed Cleaning

Bed cleaning is an important silkworm rearing process to ensure the hygiene in the immediate vicinity of silkworms in order to protect from disease infection and to ensure them good feeding appetite. Hence, timely bed cleaning is critical to maintain the worm

In a study observation on larval mortality, larval period, single cocoon weight, shell weight, length of silk thread and silk ratio were carefully noted for each treatment and replication. Three replications were used for each treatment of different eri silkworm races. Treatments were once bed cleaning frequency per day, twice bed cleaning frequency per day, and three times bed cleaning frequency per instar of Vietnamese eri-silkworm races, Indian eri silkworm races, Kenyan bivoltine silkworm races, Korean bivoltine silkworm races, and Vietnamese multivoltine silkworm races. Long spinning silk thread, robust silk cocoon and shell weight and higher percentage of silk ratio of silkworm races were registered from one time bed cleaning frequency per day and two times bed cleaning frequency per day.

The rearing tray of silkworms accumulates some unconsumed leaves after each feeding, exuviae after moulting, excreta, dead or diseased larvae, etc. All these if not cleaned, combine to form a thick and damp litter which promotes the growth of different micro-organisms, generation of heat and injurious gases and depletion of oxygen. Hence, it is very necessary to remove the litter periodically and the process of its removal is called bed cleaning.

Bed cleaning can be done by using paddy husk, straw and bed cleaning net. During 1st instar, bed cleaning should be done once per moulting, during 2nd instar twice, once after moult and before next moult. During 3rd instar thrice, i.e. after moult, before next moult and once in the middle. During 4th and 5th instars once in a day in case of shelf rearing. However, in case of floor or shoot rearing, bed cleaning should be done once in each instar.

15.2 Feeding of Silkworms

Table 15.1: Number of trays and area of space required by different instars of silkworm during rearing

	No. of trays required for 100 DFLs (diameter of tray 3.5 ft)		Space required (sq. ft) (Bi/multivoltine)	
Stage of silkworm	**Beginning**	**End of stage**	**Beginning**	**End of stage**
1st instar	2	2	4	15
2nd instar	2	5-6	15	45
3rd instar	5-6	10-12	46	90
4th instar	10-12	20	91	180
5th instar	20	40	181	360

15.2.1 During Chawki Rearing

- Distribute the worms uniformly in the trays.
- Feed the silkworm with nutritious and succulent leaf.
- Provide 3-4 times feeding with tender leaves of 0.5-2 cm cut pieces during 1st stage.
- Provide leaves @ 5 kg/100 dfls for 1st and 18 kg/100 dfls for 2nd instar worms.
- During 2nd stage cut the leaf into 2-4 cm and feed 3-4 times.

15.2.2 During Late Age Rearing

- Use mulberry leaf only from 60 days old plantation.
- Provide 4 times feeding to the silk worm.
- Maintain rearing/feeding room temperature at 25-26^{0}C and relative humidity at 65-70%.
- Use tender leaves only for resuming (only one feeding after 3rd/4th moult) then change to medium in 4th stage and coarser leaves in 5th age.
- After moulting stop feeding when 90% worms enter into moult.

15.3 Spacing

The silkworms grow very rapidly from age to age and increase many times their weight and size from the previous instar. The total increase in weight from hatching to the end of 5th instar is about 7,000 to 10,000 times. Crowded situation in rearing trays results in increased humidity, heat, fermentation of litter, all of which will in turn cause under development of larvae, wastage of feeding leaf and unhygienic condition. To provide more and adequate space for the growing worm, the rearing space has to be extended at each stage and this is called spacing. Spacing is usually done along with bed cleaning and is given once a day.

References

Central Silk Board. (n.d.). Techniques of rearing silkworm. Sericulture Information Linkages and Knowledge System. Retrieved January 19, 2025, from https://silks.csb.gov.in/anjaw/wp-content/themes/Common_District/trs-frame.html

Notes on Zoology. (n.d.). Rearing process of silkworm: 7 steps. Retrieved January 19, 2025, from https://www.notesonzoology.com/sericulture/rearing-process-of-silkworm-7-steps/219

Outcomes Assessment

Part A

Answer the following questions true or false. [1×4=4 Marks]

1. Without bed cleaning mortality is less. True/False
2. Stop feeding when 80 percent of worms stop feeding. True/False
3. Spacing is not given along with bed cleaning. True/False
4. Use mulberry leaves only after 60 days of plantation. True/False

Part B

Answer the following questions. [2×5=10 Marks]

1. Describe about the feeding of worms at late rearing stage.
2. Describe about the feeding of silkworms at chawki stage.
3. Describe about the types of bed cleaning.
4. What is bed cleaning?
5. What is the spacing given for each instar of larvae?

Part C

Write a brief note on each of the following. [5×2=10 Marks]

1. Describe briefly about the bed cleaning of worms and its advantages.
2. Describes briefly about the different types of feeding.

Part D

Select the correct answer from the four choices given under each question. [1×3=3 Marks]

1. Bed cleaning can be done by using
   a. Paddy husk
   b. Straw
   c. Bed cleaning net
   d. All of the above
2. Late Age Rearing requires how many times feeding
   a. 3 times
   b. 4 times
   c. 5 times
   d. 6 times
3. Process of removal of litter periodically is called as
   a. Bed cleaning
   b. Spacing
   c. Moulting
   d. None of these

Answer Key: 1. **d,** 2. **b,** 3. **a.**

16

Identification and Picking of Ripe Silkworms

16.1 Process of Ripening of Worms

Closer to the end of the fifth stage, the silkworm stops ingesting and turns into the so-called mature larva and starts spinning the cocoon. Properly reared worms, when ripe, will attempt to spin true cocoons even below incorrect conditions. However, it needs to be remembered that the worm spinning the cocoon is to defend itself from outside disturbances and enemies all through the maximum crucial period of its metamorphosis. The rearer can get the reward for his labour most effective, if he cooperates by making the situations such that, through spinning a commercially good cocoon it also most correctly protects itself. This cooperation is composed in imparting the ripe silkworm with appropriate mountages to facilitate its spinning of suitable cocoons.

The eating duration throughout the fifth instar may close for 5 to 7 days in the case of multivoltine and bivoltine races in the tropical regions and 7 to 9 days in the case of bivoltine and univoltine races in sub-tropical areas. The mature worm is effectively distinguishable by its translucent colour. The frame also shrinks with duration; there is also a constriction seen on the 4th and 5th segments. The worm loses appetite and begins to look approximately as although on the lookout for a appropriate area to which to attach itself. Generally, the ripening worms circulate in the direction of the periphery of the rearing trays in search of anchorage to begin spinning of cocoons. This is the time for selecting the ripe worms and putting them on mountages. This process is known as 'mounting'. It has to be remembered that all the worms in a tray do not ripen simultaneously. It commonly happens that by the time all the worms in a tray are ripe, a number of the early worms can be over-ripe. They could have misplaced a quantity of silk by the point they are picked up, and they are furthermore in a hurry to spin their cocoons that they continue to do it anyways, without ready to select a right location, and as a consequence produce irregularly formed and flattened or sticky cocoons causing loss to the rearer. If the worms are established earlier than they are fully ripe, they will flow about and every so often fall off, and also they defecate in the cocoonage, frequently soiling and reducing the price of the cocoons which different worms are spinning. The cocoons spun by the worms which might be established too early could be smaller and inferior. It is, therefore, of paramount importance that the rearer should be ready with the desired quantity of mountages properly in time and also mount the worms while they are completely mature. That is, the time when the rearer should not hesitate to rent enough quantity of people to pick out the ripe worms and mount them. Any negligence at this stage is sure to price him heavily.

16.2 Process of Spinning

As the mature worms are mounted on the mountages, their last excreta are semisolid. When it rains and the humidity is high, extra bodily moisture is expelled as liquid pee during the mounting process. After defecation, the worm begins spinning its cocoon. It first attaches itself to the mountage by leaking a little droplet of the silk fluid, which instantly solidifies and adheres to it. The silkworm then takes silk fluid from the two silk glands on either side of its body by repeatedly swinging the anterior half of its body. Silk fluid is secreted in minute amounts and hardens to create the long, continuous silk strand. At first, the worm weaves a loose hammock, which gives it with the essential foothold to begin spinning the cocoon properly. The filament is spun in the shape of or, with the former being frequent in the outer layers of cocoon shells and the latter in the middle and interior layers. Layers upon layers of filament are laid, forming the Cocoon's tight shell. The worms spin the cocoon over a period of 1 to 2 days for multivoltines and 2 to 3 days for uni/bivoltines. During this phase, the silkworm larvae must be kept undisturbed because shaking during cocoon spinning causes spinning to stop and even breaks the thread. It is also critical to provide adequate ventilation since the worms must expend a significant amount of moisture throughout the spinning process; the silk, while solid, is still wet and must dry before it can form a stable cocoon.

References

Unknown. (n.d.). Mounting and Spinning. https://hbmahesh.weebly.com/uploads/3/4/2/2/3422804/spinning_and_harvesting.pdf

Part A

Answer the following questions true or false. [1×4=4 Marks]

1. Mature worms are not translucent. True/False
2. From 2nd instar onwards worm stops feeding. True/False

3. Eating period in 5th instar lasts for 2 days. True/False
4. Mounting too early leads to smaller and inferior worms. True/False

Part B

Answer the following questions. [2×3=6 Marks]

1. Describe the stages of silkworm up to maturing and features of matured worm.
2. Describe about the picking of silkworms.
3. Precautions to be taken in selection of ripe worms.

Part C

Write a brief note on each of the following. [5×2=10 Marks]

1. Describe briefly about the ripening, identification and picking of ripe worms.
2. Describe briefly about the process of spinning of ripe worms.

Part D

Select the correct answer from the four choices given under each question. [1×3=3 Marks]

1.The process for selecting the ripe worms and putting them on mountages is called as

a. mounting b. spacing
c. cleaning d. all of these

2. For multivoltines races they spin the cocoon over a period of

a. 1 to 5 days b. 1 to 2 days
c. 1 to 4 days d. 1 to 3 days

3. For bivoltines races they spin the cocoon over a period of

a. 2 to 5 days b. 2 to 4 days
c. 2 to 3 days d. 1 to 3 days

Answer Key: 1. **a,** 2. **b,** 3. **c.**

17

Mounting and Types of Mountages

17.1 Introduction

The most important device that helps or supports the silkworms (larvae) for comfortable spinning their cocoon is called cocoonage or mountage. It determines both the quality and quantity of the cocoons. Different types of cocoonages are used in different parts of India. In general, these are made of wood, bamboo, cardboard, plastic, grass, dry leaves, twigs, etc.

Mounting is an essential step in the rearing of silkworm. Mounting indicates the transition of silkworm from larval stage to pupal stage where it prepares to spin cocoons. Mounting is a process where fully grown (specifically 5th instar) silkworm larvae are picked and mounted in a suitable place (mountage) where it is about to spin. Mounting is a pivotal phase in silkworm rearing.

The following are the objectives of mounting:

1. To provide an ideal space for silkworm larvae to spin,
2. To ensure good quality cocoons,
3. To minimize mortality among silkworm by avoiding overcrowding,
4. To ensure a maximum silk production.

An ideal mountage should fulfill the following requirements:

1. Convenient space with suitable dimension for spinning proper-sized cocoon.
2. Should not allow formation of double or malformed or flimsy cocoons.
3. Should have provisions for cleaning the excreta of spinning larvae.
4. Should be suitable for easy harvesting of cocoons.
5. Should be cheap, durable and easy to handle.

17.2 Mounting Tips

1. **Mounting time:** Mounting can be performed both during day and night time with optimum temperature and humidity attained. As soon as larvae attain maturity consider the larvae for mounting without any delay.

2. **Mounting materials:** Mounting material is an important factor for successful silkworm rearing. Choosing a correct type of mounting material helps the silkworm to produce good quality silk. Various types of mounting materials such as bamboo, plant shoot, paddy straw, plastic brush are used. In recent studies it is also found that mango twigs of various lengths can be used as a mounting material.

3. **Mounting chamber/room:** The mounting room or chamber should be well ventilated and located in ideal shaded area. Various climatic factors like temperature, humidity, wind speed and light should be well maintained.

4. **Mounting density:** While transferring mature larvae to a mounting material, it is recommended to maintain optimum density of worms per metre square. Optimum density recommendation is 800-900 worms/m^2. Its failure leads to overcrowding of silkworm larvae which ultimately leads to faulty cocoon formation and low silk yield.

5. **Repellents:** Repellents act as guide to silkworm to the spinning area. They are used during mounting process to prevent the larvae from spinning in unwanted area of the mounting material. Various indigenous substances such as lime powder, turmeric powder and neem leaves are used as repellents.

Mounting

- Apply Sampoorna @ 20 ml (dissolved in 4 litres of water) per 100 dfls over the leaves for early and uniform spinning of cocoons.
- After attaining full growth in the final instar, the worms cease to eat and are ready to spin.
- Such worms are slightly translucent and raise their heads to find a place for spinning.
- These worms have to be picked up and transferred to a mountage for spinning cocoons.
- Mounting of worms should not be delayed as the ripened worms will waste silk.

- About 800-900 worms per m^2 are to be kept on a mountage. For 100 dfls, about 30 to 40 chandrakis are required.
- Mountages should be kept under shade in well-ventilated place.

17.3 Mounting Method

There are mainly 3 types of mounting method generally practiced in India; they are as follows:

1. Picking up and mounting
2. Self-mounting method
3. Shoot shaking method

1. **Picking up and mounting**: This method is commonly used by south Indian farmers. It is generally performed by picking individual matured larvae by hand only and mounted on mounting frame. This method helps to produce good quality cocoon as while hand mounting, worker select larvae which is disease free and place them at a proper spacing in mounting frame. It is very time-consuming process (**Fig.17.1, Fig.17.2**).
2. **Self-mounting:** It is also called natural mounting as mature silkworm crawl up into the cocooning frame placed on the rearing bed. The cocooning frames are left on the rearing bed at about 4-6 hours to allow the larvae to crawl up into the frame (**Fig.17.3**).
3. **Shoot shaking mounting:** When more than 30% of worms are matured, 10-15 number of mulberry shoots are picked from the feeding rack and shaken gently so that the matured larvae fall down on mounting frame which is placed on the floor. It is a time saving process.

Fig. 17.1: Mature silkworms

Fig. 17.2: Picking of mature silkworms

Fig. 17.3: Self-mounting method

17.4 Types of Mountages

The most significant device that assists or supports mature silkworm larvae in freely spinning their cocoon is known as mountage or cocoonage. Mountages are crucial in determining the quality and number of cocoons produced. Mountages are made up of various things such as bamboo, plant shoot, paddy straw, plastic brush etc.

There are various types of mountages used in rearing of silkworm:

1. Rotary card board mountage
2. Bamboo *chandrika*
3. Plastic collapsible mountage
4. Bottle brush mountage
5. Bamboo spiral mountage
6. Bamboo screen mountage
7. Shoot rearing rack mountage

1. Rotary card board mountage: Rotary card board mountage is best among all mountage because of its uniform shape and size of cocoon produced. This type of mountage is most popular in Japan. The size of rotary mountage is 55×60 cm^2. Its each set of board can accommodate 1560 number of larvae (**Fig. 17.4**).

Fig. 17.4: Rotary mountage with spinning larvae

Advantages

- Uniform shape and size.
- Less number of defective cocoon is produced.
- Easy harvesting.
- Easy storage and disinfection.
- Higher reliability and improved reeling parameters.

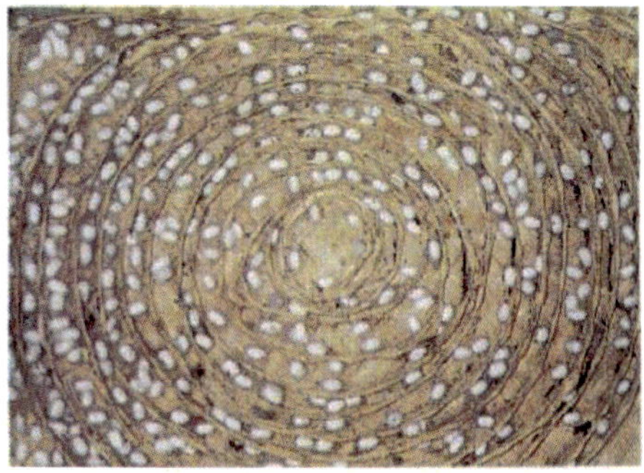

Fig. 17.5: Bamboo chandrika

2. Bamboo *chandrika*: These are the most common type of mountage used in India (**Fig.17.5**).

Characteristics

- Made of up bamboo spiral woven on a bamboo mat with 2 supports made of bamboo sticks.
- 1.2-1.8 metre in size.
- Spirals are made of mats of size 5-6 cm width.
- Small holes are made on the mat base to provide ventilation.
- Durability is up to 4 years.
- These are cheap and easily available.
- Can accommodate 900-1000 worms.

Advantages

- Easy availability.
- Absorbs moisture easily.

Disadvantages

- Falling of dead and diseased larvae on the floor leads to contamination.
- Separate space required for mounting and storage of mountage.
- Due to sagging of spirals, uneven space for cocoon construction.
- More space required for storage.
- Time and labour consuming for mounting.
- Difficulty in handling of mountage.
- Less life span (about 3-4 years)

3. Plastic collapsible mountage: It is made by plastic mesh having nearly 11 folds of 2.2 inch in height and can be placed in a wooden tray of 2×3 feet size for mounting. It can accommodate 350-400 numbers of larvae. It is also called Netrika (**Fig.17.6**).

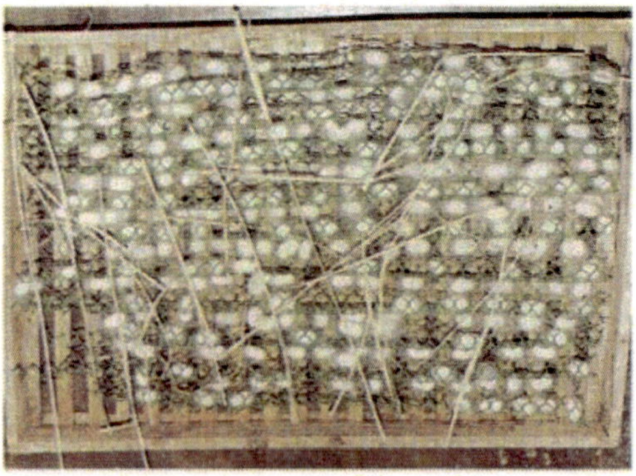

Fig. 17.6: Plastic collapsible mountage

Advantages

- Can be preserved by folding when not in use.
- Easy for handling and disinfection.
- Does not require additional space for mounting.
- Durable for long time.
- Lighter in weight.

Disadvantages

- Loses size, shape and corrugated easily.
- Easily damaged by rodents.

4. Bottle brush mountage: It is machine made plastic material with individual detachable pieces. Nearly 50-60 individual pieces are joined together by an iron rod and at the end pole it has an iron stopper. The surface of plastic material is granulated to provide grip for silkworm spinning. Nearly 350-400 number of larvae can be accommodated by this type of mountage (**Fig.17.7, Fig.17.8**).

Advantages

- Easy disinfected.
- Durable for long time.
- Detachable when not in use.

Disadvantage

- Cannot be made locally when damaged.

Fig. 17.7: Plastic bottle brush mountage

Fig. 17.8: Bamboo bottle brush mountage

5. Bamboo spiral mountage: It consists of 2 bamboo supports of length 100 cm and on it bamboo net is woven as tapes at a distance of 5 cm spirally. When the larvae are ready for mounting, they are picked up from rearing tray, old newspaper is placed on the tray and bamboo spiral structure is inserted into the rearing tray and the mature silkworm are mounted. 300 number of silkworms can be mounted in a round bamboo astray of 90 cm diameter (**Fig.17.9**).

Fig. 17.9: Bamboo spiral mountage

Advantages

- Easy for mounting.
- Requires less labour and space.

6. Bamboo screen mountage: It is made up of bamboo sticks arranged one above the other by cotton thread leaving a space of 3 cm which can provide space for silkworm spinning. The support provides anchor to the spinning worms. When the worms start maturing, they are picked up and placed on the mountage slowly. 50 worms are mounted in an area of 1sq foot and after mounting, it is hung from the roof.

Advantages

- More durable as compared to bamboo spiral mountage.
- Requires less space in a room.
- This type of mountage provides good aeration to larvae while spinning.

7. Shoot rearing rack mountage: It is used to facilitate accommodation of mountages within the shoot rearing rack itself. It saves space and labour requirement.

References

Sericulture::SilkwormLateageRearing.(n.d.).https://agritech.tnau.ac.in/sericulture/seri_silkworm4_lateage%20rearing.html#:~:text=Mounting%20of%20worms%20should%20not,shade%20in%20well%20ventilated%20place.

Kadam, S. (2016, July 21). *Mountage: Meaning and Types | Sericulture*. Zoology Notes. https://www.notesonzoology.com/sericulture/mountage-meaning-and-types-sericulture/20

Outcomes Assessment

Part A

Answer the following questions true or false. [1×5=5 Marks]

1. Mounting indicates the transition of silkworm from larvae to adult stage. True/False
2. Mounting is done when the larvae is in state of 5th instar. True/False
3. In shoot shaking method at least 50% of the larvae should be matured for mounting. True/False
4. Rotary card board mountage is vastly used in Japan. True/False
5. Bamboo chandrika is also called Netrika. True/False

Part B

Answer the following questions. [2×5=10 Marks]

1. Define Mounting and its objectives?
2. Briefly explain the ideal environmental condition required for mounting?
3. Why mounting density plays a crucial role in rearing of silkworm?
4. Describe repellent and its benefits in mounting?
5. Why mounting of larvae is necessary when they attain maturity without any delay? Write the consequences of delaying.

Part C

Write a brief note on each of the following. [5×5=25 Marks]

1. Briefly describe the mounting process and its safety precautions.
2. Briefly describe the different methods of mounting and evaluate why self-mounting is also called as natural mounting.

3. Briefly describe the type of mountages used in silkworm rearing.
4. Briefly describe the mostly used mountage in India. Write about the advantages and disadvantages.
5. Describe the need and importance of mounting materials used in silkworm rearing.

Part D

Select the correct answer from the four choices given under each question. [1×10=10 Marks]

1. Mounting is done at which larval instar?
   a. First b. Second
   c. Fourth d. Fifth
2. Rotary card board mountage is vastly used in which country?
   a. India b. China
   c. Japan d. Brazil
3. The optimum population density required for silkworm rearing is about
   a. 400-500 worms/ m^2 b. 800-900 worms/ m^2
   c. 1000 worms/m^2 d. No minimum requirement.
4. Which cannot be used as a repellent in silkworm rearing?
   a. Turmeric powder b. Maleic Hydrazide
   c. Neem leaves d. Lime powder
5. Plastic collapsible mountage is also called as
   a. Rowrike b. Supernet
   c. Netrika d. All of the above
6. An iron rod used as a central material in bottle brush mountage that act as
   a. Supporting material b. Grip provider
   c. Rearing place d. All of the above
7. Major disadvantage for not using mulberry twig mountage is
   a. Double coccon formation b. No coccon formation
   c. Piercedcoccon formation d. Stained coccon formation
8. Thalaghattapura ribbon chandrikeis associated with which state of India
   a. Bihar b. Karnataka
   c. Jharkhand d. Odisha

9. Vertical fixed type mountage is a
    a. Hybrid mountage
    b. Low cost mountage
    c. Simple mountage
    d. All of the above
10. The mostly used mountage in India can accommodate how many worms in itself
    a. About 2000 worms
    b. About 500 worms
    c. About 1000 worms
    d. About 5000 worms

Answer Key: 1. **d,** 2. **c,** 3. **b,** 4. **b,** 5. **c,** 6. **a,** 7. **a,** 8. **b,** 9. **a,** 10. **c.**

18

Cocoon Harvesting and Grading

18.1 Introduction

The cocoons are harvested and sold to the reeling agencies. Harvesting has to be done after the sclerotization of the pupal cuticle and before the emergence of adult moth. It has been recommended to harvest at the fifth day of spinning for tropical races and seventh or eighth day for temperate races. Delay in harvesting leads to pierced cocoon by the emerging moth and early harvesting causes injury to the unsclerotized pupa and thereby soiling the cocoon. The cocoons are normally harvested by hand or by simple devices. Faecal pellets if any on them must be removed and cleaned. Healthy cocoons must be separated from the defective ones and sorted out based on their size.

18.2 Harvesting of Cocoon

After completion of spinning of cocoon, the larval skin is cast off and pupation takes place. This generally occurs on the 3rd/4th day of spinning in the case of multivoltines and 4th/5th day of spinning in bivoltines and univoltines in the temperate regions. The pupa when formed has a thin cuticular skin which is soft to touch and may get ruptured easily, if disturbed. Too much delay in harvest is also not desirable, as it reduces the weight of cocoons. Harvesting of cocoons is normally done by hand. Only in the case of revolving mountages simple devices are used. The harvesting process is the best time for sorting cocoons according to the quality as good cocoons are easily distinguishable from double and defective cocoons.

- Multi-voltines: Spinning completes on 3rd to 4th day - harvesting on 5th day.
- Bivoltines Spinning completes on 4th to 5th day - harvesting on 6th to 7th day.

General properties of silk

1. Silk is crystalline
2. Homogenous in structure

3. Hygroscopic in nature
4. Light in weight
5. Longest and strongest of all natural fibres
6. Soft, lustrous and hygienic
7. Excellent affinity for dyes - takes colours easily
8. Does not catch fire easily/quickly as nylon or wool
9. It is elastic and has elongation of 20%
10. Has high tensile strength, breaking strength is 4 g per denier.

18.3 Parameters for Classification or Grading of Cocoon

1. Reelability ratio $= \dfrac{\text{No. of cocoons reeled}}{\text{No. of feeding ends}}$

2. Reelability percentage $= \dfrac{\text{No. of cocoons reeled}}{\text{No. of feeding ends}} \times 100$

   Where, No. of cocoons reeled = number of sample cocoons – number of unreelable new cocoons + number of converted carry-over cocoons

   No. of feeding ends = number of cocoons fed + number of carry-over cocoons – number of converted carry-over cocoons

3. Floss percentage $= \dfrac{\text{Weight of floss (from 50 shells)}}{\text{Weight of 50 cocoons}} \times 100$

4. Denier $= \dfrac{\text{Total Wt. of reeled silk(g)}}{\text{Total length of reeled silk(m)}} \times 9000$

5. Raw silk percentage $= \dfrac{\text{Weight of silk + carryover cocoon silk yield}}{\text{Weight of cocoons used}} \times 100$

6. Shell ratio % $= \dfrac{\text{Single shell weight (g)}}{\text{single cocoon weight (g)}} \times 100$

7. Procedure for estimation of Renditta:

   The procedure has certain constants for estimating the renditta from the shell ratio which are as follows:

   1. 165 for cocoons with shell ratio of 14-16%
   2. 150 for cocoons with shell ratio of 17-20%

3. 133 for cocoons with shell ratio of 21-23%

The Renditta is estimated as = Constant / Shell ratio

8. $$\text{Loss (\%) on mountage} = \frac{\text{No. of larvae mounted} - \text{No. of cocoons harvested}}{\text{No. of larvae mounted}} \times 100$$

9. After cocoon sorting, the percentage of good and defective cocoons are calculated. This gives reelable cocoon percentages. All the test values help in estimating the quality of raw silk of that particular lot. The percentages of all defective cocoons and good cocoons are calculated individually by number and weight using the following formulae:

$$\text{\% of total defective cocoons} = \frac{\text{No. of defective cocoons}}{\text{Total no. of cocoons}} \times 100$$

$$\text{\% of total good cocoons} = \frac{\text{Weight of good cocoons}}{\text{Weight of total cocoons}} \times 100$$

18.4 Grading of Cocoon

In India, the grading of cocoons is primarily conducted according to various criteria such as size, shape, weight, colour, and silk quality. The classifications may differ slightly based on the type of silk (mulberry, tasar, eri, muga) and the specific practices of different regions. The following are the general categories:

1. Mulberry Cocoons (Silkworm *Bombyx mori*)

Grade A: Large, uniform cocoons that are white or golden-yellow, characterized by their density and smoothness.

Grade B: Medium-sized cocoons with a somewhat uneven shape, possessing good colour and a smooth surface.

Grade C: Smaller cocoons that may be slightly thin or irregular, yet yield an acceptable amount of silk.

Grade D: Very small and irregular cocoons, typically associated with a lower silk yield.

2. Tasar Cocoons (Wild Silkworm *Antheraea mylitta*):

Grade A: Larger, well-formed cocoons that are uniform in shape and exhibit a shiny brown hue.

Grade B: Medium-sized cocoons that display minor irregularities in both colour and shape.

Grade C: Smaller cocoons that may appear dull or exhibit slight damage.

3. Eri Cocoons (Silkworm *Samia ricini*)

Grade A: White, dense, and uniform cocoons that are oval or spindle-shaped.

Grade B: Slightly smaller cocoons that lack uniformity.

Grade C: Small, thin, or slightly defective cocoons.

4. Muga Cocoons (Silkworm *Antheraea assamensis*):

Grade A: Large, golden-yellow cocoons that are oval-shaped and dense.

Grade B: Medium-sized cocoons with minor irregularities in shape and texture.

Grade C: Smaller cocoons that may have a dull appearance or thinner shells.

Additional Grading Parameters

1. Defective Cocoons: This category includes cocoons that are pierced, double, or flimsy.
2. Reeling Performance: The grading process also takes into account the reeling capacity (silk yield) and the length of the filaments.

Specific numerical grades ranging from 1 to 10 may be assigned based on regional practices or for auction purposes, although these are generally derived from the afore-mentioned criteria. The precise grading systems can vary according to local standards and customs.

References

Central Sericultural Research & Training Institute, Mysuru. (n.d.). A guide for silkworm rearing. Retrieved December 19, 2024, from http://www.csrtimys.res.in/sites/default/files/phamplets/en-00-guide.pdf

Food and Agriculture Organization of the United Nations. (n.d.). Silk reeling and testing manual. Retrieved December 19, 2024, from http://ftpmirror.your.org/pub/misc/cd3wd/1005/ag_silk_reeling_testing_unfao_en_lp_116630.pdf

Outcomes Assessment

Part A

Answer the following questions true or false. [1×4=4 Marks]

1. Tasar silkworm having five types of grades. True/False
2. Multivoltine worms take six days for spinning. True/False
3. Silk should have an elongation of forty percent. True/False
4. Univoltine worms take 4 to 5 days for spinning. True/False

Part B

Answer the following questions. [2×5=10 Marks]

1. Describe about the grading of mulberry cocoons.
2. Describe about the grading of tasar cocoons.
3. Describe about the harvesting of cocoons.
4. What is Floss percentage?
5. What is reelability ratio?

Part C

Write a brief note on each of the following. [5×2=10 Marks]

1. Describe briefly About the methods of grading up cocoons.
2. Describe briefly about the grading of different types of cocooon.

Part D

Select the correct answer from the four choices given under each question. [1×3=3 Marks]

1. The cocoons that are pierced, double or flimsy are called as
   a. Double cocoon
   b. Spoiled cocoon
   c. Pierced cocoon
   d. All of the above
2. Silk is ____________ in structure.
   a. Hexagonal
   b. Crystalline
   c. Homogeneous
   d. None of the above
3. Silk has _________________ strength.
   a. High tensile strength
   b. Low tensile strength
   c. Medium tensile strength
   d. All of the above

Answer Key: 1. **a,** 2. **c,** 3. **a.**

19

Diseases and Pests of Mulberry Silkworm

19.1 Introduction

Mulberry silkworm, *Bombyx mori,* is affected by a number of diseases caused by viruses, bacteria, fungi and protozoa. These illnesses are recognized to occur in practically every region where silkworms are raised in the world seriously harming the harvest of silkworm cocoons. Several actions have been proposed as a means of preventing and managing the illnesses, but none of them had shown signs of foolproof with the outcome that one must consistently be cautious to eradicate the core infection's source in addition to avoiding cross-infection. Additionally, care has to be taken in order to ensure that they are not exposed to stressful circumstances such as low temperature and inadequate nourishment, poor ventilation, and humidity which could lead them easily susceptible to various diseases.

19.2 Diseases of Mulberry Silkworm and Their Control

19.2.1 Protozoan Diseases

The protozoa that harm silkworms are the parasitic ones from the genera *Nosema, Pleistophora,* and *Thelohania* and class Microsporidia. In addition to this, reports of *Leptomonas* and *Trypanosoma* infections in the silkworm *Bombyx mori* L. have been made. However, the pebrine disease is the main protozoan illness of the silkworm. It is also known as 'pepper disease' because of the development of areas of black pepper after infection.

19.2.1.1 Pebrine Disease

The devastating and long-lasting pebrine illness affects *Bombyx mori* L., the silkworm. The rapid collapse of the silkworm industry in France and Italy in 1965 was caused by this disease. Although this disease has been fought for more than a century in all sericultural countries, it has not yet been eradicated. Nonetheless, it has been contained by adhering to the methods of stringent mother moth inspection for the provision of disease-free silkworm eggs, as well as cleaning and sanitary rearings. The disease is under reasonable control, but because of infected seed and ongoing secondary contamination in the rearing house, it nevertheless occasionally recurs.

Cause of the disease

Pebrine is caused by *Nosema bombycis* Nageli belonging to family Nosematidae of order Microsporidia. The pathogen infects the host through feeding of contaminated mulberry leaf (peros) and also by rearing infected silkworm eggs (transovarial). There are a lot of different sources of infection. The rearing of transovarialy and surface contaminated layings is the primary source. Diseased and dead larvae, their excrement, moths, contaminated egg shells, exuviae from larvae and pupae, etc. can also cause infection. Dust from infected rearing and leaf storage rooms, contaminated trays, seat paper, and diseased larvae's excrement are the main sources of infection in the rearing bed. Infection can occasionally occur from a mulberry leaf that has been infected in the field. Another source of infection could be dead larvae and excrement from wild insects infected with pebrine.

Symptoms

The symptoms of this disease can be observed in all the stages of silkworm, *viz*., egg, larvae, pupa and adult. These symptoms form important criteria for identifying the disease. In the egg stage, poor number of eggs laid, lack of adequate sticking of eggs to the substratum, lack of egg uniformity, a greater number of unfertilized and dead eggs, poor and irregular hatching are some of the symptoms. Sometimes infected eggs cannot hatch out and hatched larvae may also die. Larvae show poor appetite, retarded growth and development leading to non-uniformity in size, moulting is irregular and larve become sluggish. Transovarial infected larvae die before third moult but those which are heavily infected die during first instar itself. The larva usually shows

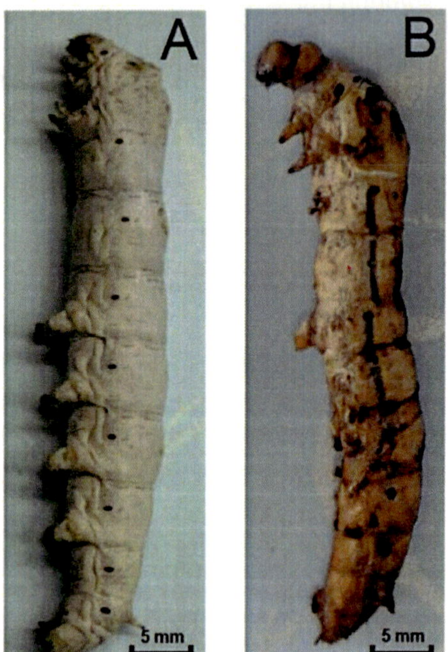

Fig. 19.1: Pebrine disease

wrinkled skin with rustic brown colour and in the moribund stage they do not rot but remain rubbery. The affected gut becomes opaque and the silk gland shows white pustules in different places along its length. Sometimes black irregular pepper like spots are noticed on larval skin. The infected pupae are flabby and swollen with lustreless and softened abdomen. Sometimes irregular black spots are noticed near the rudiments of the wing and abdominal area. Highly infected pupae fail to metamorphose into adults. The moth emergence is delayed and improper. They have clubbed wings with distorted antennae and do not mate properly. The scales from wings and abdominal area easily come off. In infected moths if the accessory glands are infected, the moth may lay eggs with less gluey substance resulting in their detachment from the egg cards (**Fig.19.1**).

Prevention and Control

The fundamental measure for the prevention and control of this disease is to produce healthy eggs, so as to avoid embryonic infection, conducting systematic mother moth examination for producing healthy eggs. The other methods include thorough effective disinfection of rearing rooms, equipment and surroundings, and maintenance of strict hygienic conditions during rearing. It is essential to surface disinfect the laying in 2 % formalin solution for 10 minutes before incubation. If the eggs are in advanced stage of embryonic development surface disinfection is done with 1 % formalin for 5 minutes. Young silkworms should be reared under hygienic conditions. As a precaution test examination of unhatched blue eggs, dead eggs, hatched larvae and eggshells can be done and if pebrine is detected, such eggs should not be brushed and if brushed the larvae should be destroyed. Eggs after surface disinfection should be dried and stored in a separate room away from egg production and examination room. To minimize the incidence of pebrine immersion of the silkworm eggs in hot water, high temperature treatment of the pupae, dipping of the eggs in hot hydrochloric acid is effective. Preventive methods like chemotherapy of Nosema infection has been reported through a number of antimicrosporidian drugs like fumagillin, benomyl, bengard, bavistin, ethyl and methyl thiophanate but preventive methods have always been found to be better than the curative measures.

19.2.2 Bacterial Diseases/ Flacherie Disease

Bacterial diseases affecting silkworm are collectively known as flacherie due to the flaccid nature of the diseased larvae. The incidence of flacherie is high during hot and humid weather. In general, massive out-break of these diseases are uncommon but depending upon poor disinfection, accumulation of faeces in the rearing trays, feeding of mulberry leaves with contamination, improper handling and unsafe use of bacterial pesticides, cause huge loss.

Cause of the disease

This disease is caused by the multiplication of a large number of bacteria, Bacilli, Streptococci and Staphylococci in the haemolymph. The route of infection is through injury or wounds and rarely perorally. Two major types of bacterial septicemia are generally observed. One is the black thorax septicemia caused by *Bacillus* sp. belonging to the family Bacillaceae of the order Eubacteriales, size 1-1.5 × 3 microns, spores subterminal, gram-positive in nature and the other is the red septicemia or Serratia septicemia caused by the bacillus *Serratia marcescens,* size 0.6-1.0 × 0.5 microns, non-sporulating and gram-negative.

Symptoms

Larvae show symptoms like sluggish movement, decreased appetite, straightened body, swollen thorax, shrinkage of abdominal segments, vomiting and bead like faeces and loss of clasping power of legs. Further, the body becomes soft and discoloured and the body wall ruptures easily emitting foul smelling fluid spreading quickly. At higher temperature and under epidemic conditions they may die within 5-6 hours (**Fig.19.2**).

Fig. 19.2: Flacherie diseased silkworms dead and putrefied

Prevention and control

The bacterial agents that cause these illnesses thrive in environments with high temperatures and humidity, that is why these illnesses are primarily associated with these seasons. Usually, they come after an infection of the wound. The bacteria typically enter through the wound and grow in the haemolymph, interfering with normal physiological processes and leading to septicaemia.

Since the larvae in their fifth instar are more vulnerable to harm, these illnesses typically manifest later in this stage. Keeping the raising chamber, equipment, and mulberry leaves clean can help prevent the growth of these bacteria, which can be an effective way to control these diseases. It is important to take precautions to prevent harming the worms, packing trays too full and accumulation of faeces in the rearing bed.

19.2.3 Viral Diseases / Grasserie

Since viruses cause about 70% of all disease-related losses in silkworms, they are a serious threat to the industry. There are inclusion and non-inclusion forms of silkworm viruses. Typical inclusion bodies are made up of inclusion virus infections. These are more easily recognized by standard microscopy as nuclear polyhedrosis and cytoplasmic polyhedrosis. The non-inclusion form includes Densonucleosis and infectious flacherie, which are only detectable by serological testing and electron/fluorescent microscopy.

19.2.3.1 Nuclear Polyhedrosis

It is a year-round virus disease that is among the most dangerous in tropical nations. Other names for this illness are hanging sickness, fatty degeneration, jaundice, grasserie (**Fig.19.3**) and milky disease.

Reasons behind the illness: The *Borrelina bombycis* virus, which is a member of the Baculoviridae subgroup A, is the cause of this illness. As the name suggests, this virus divides and creates polyhedral in the nuclei of blood, adipose tissue, cutaneous and tracheal epithelial cells. On rare occasions, the nuclei of the silk gland cells' posterior and middle portions are also affected. The size of the rod-shaped virus particles is approximately 330 × 80 nm. Typically, the shape is octahedral or hexahedral but occasionally tetragon.

Infection occurs mostly through feeding on mulberry leaves contaminated with polyhedral. Wounds are rarely the site of infection. This disease has also been found to be induced by heat, cold and chemical treatments. High temperatures and humidity, together with their abrupt swings, inadequate ventilation in the rearing room, inadequate sanitation of the equipment and the rearing room and feeding of sensitive leaves during late instars

Fig.19.3: Grasserie disease

are all contributing factors to the disease outbreak. The emergence and spread of the disease have also been linked to inadequate larval spacing, hunger and excessive dampness in the rearing bed.

Symptoms

Other than the worms being a little lethargic, there are no signs in the early stages of the illness. At first, the skin seems shiny and oily. The body turns milky white with intersegmental swellings and thin, brittle skin as the disease progresses. The delicate skin is brittle and quickly breaks, releasing liquid bodily fluids that contain a myriad of polyhedral and serving as an additional source of infection. Another defining feature of this illness is that the larvae get restless and wander aimlessly over the rims or ridges of rearing trays, eventually collapsing to the ground and dying. After infection, young larvae die after 4-5 days, while adult larvae die after 5-7 days. With the exception of the caudal legs, which are used to hang the head downward, diseased larvae lose their ability to clasp together. When an infection occurs too late, the worms are able to spin the cocoons but die inside, resulting in melted cocoons. If the infection occurs too early, the worms are unable to create the cocoons and die.

Prevention and control

The mounting rooms, mulberry storage rooms, silkworm rearing rooms, equipment, and rearing premises should all be well cleaned and disinfected before brushing in order to effectively avoid this disease. Essentially, the eggs should only have their surfaces cleaned. Rearing silkworms requires very tight hygiene guidelines. With the greatest amount of new polyhedral available, the ill and dead larvae provide the main source of infection throughout raising. As a result, the unhealthy larvae should be carefully removed without rupturing the skin, and they should be burned or placed in lime vats for proper disposal. The ideal humidity and temperature should be given to the silkworms based on their stage. Cross ventilation should be used to ensure fresh air circulation during the IV and V instars. Mulberry leaves, which are rich in nutrients, should be fed to the silkworms during later stages. Feeding of tender leaf should be avoided. With the greatest amount of new polyhedral available, the ill and dead larvae provide the main source of infection throughout raising. As a result, the unhealthy larvae should be carefully removed without rupturing the skin, and they should be burned or placed in lime vats for proper disposal. The ideal humidity and temperature should be given to the silkworms based on their stage. Cross ventilation should be used to ensure fresh air circulation during the IV and V instars. Mulberry leaves, which are high in nutrients, should be offered to the silkworms; later on, feeding them delicate leaves should be avoided. The ideal leaf spacing and quantity should be provided based on the stage of the larvae.

Using specific bed disinfectants may help stop the disease from spreading and causing secondary contamination in addition to the previously mentioned measures. Because paraformaldehyde compounds are recognized to have anti-microbial qualities, many formulations containing this substance have been created, such as Reshamkeet Oushadh in India and Papazol in Japan. 1% captan (N-Trichloromythyl Thio-4-Cyclohexane 1,2-Dicarboxymide), 1% paraformaldehyde (Trioxymethylene), 2% benzoic acid, and 96% slaked lime powder are the ingredients in the latter bed disinfectant formulation, which provides dual protection against grasserie and muscardine. With the use of a thin cloth, dust should be applied to the larvae and bed @ 2-3 g per 0.1 square metre during early instars and 4-5 g per 0.1 square metre during IV and V instars. The dusting should be done preferably once after each moult, half an hour before resumption of feed. An additional dusting should be done on the 4th day of final instar after bed cleaning. The dusting should not be done when the larvae are under moult or preparing for moult. The quantity of Reshamkeet Oushadh required for 100 disease-free layings (40,000 larvae) is between 3-3.5 kg.

19.2.4 Fungal Diseases

A few parasitic fungi cause fungal infections in silkworms, which are also known as mycosis. Aspergillosis and mycocarditis are the two main types of these diseases. Muscardine comes in a variety of forms, and it has been dubbed white, green, yellow, black, red, and other variations based on the colour of the spores that cover the silkworm's body and give it its distinct colour. However, white and green muscardine are the more prevalent muscardine illnesses. Furthermore, it is discovered that Aspergillosis also occurs. Muskine sickness is also known as calcino because the fungal illness that attacks the silkworm eventually causes it to become hard and chalky.

19.2.4.1 White Muscardine

It is the most common and widely prevalent fungal disease found in all sericultural countries. This disease occurs usually during rainy and winter seasons under moderate to low temperature and high humidity conditions.

Causes of the disease

Several species of *Beauveria* are responsible for this illness, with *Beauveria bassiana* (Balsamo) Vuillemin (**Fig.19.4**) being the most virulent. This fungus is a member of the class Fungi Imperfecti, family Moniliaceae, order

Moniliales. Body contact is the primary method of infection; wounds and ingestion are less common. Mummified larvae, contaminated seat paper, trays, and dead wild lepidopterous larvae from the mulberry field are the main sources of infection. Because the conidia are spread by air, the disease is extremely contagious. *Beauveria bassiana* has three distinct stages in its developmental cycle: the conidium, vegetative mycelium, and aerial mycelium. When collected in a mass, the conidium is porcelain white, globular, or very rarely oval in shape, and colourless.

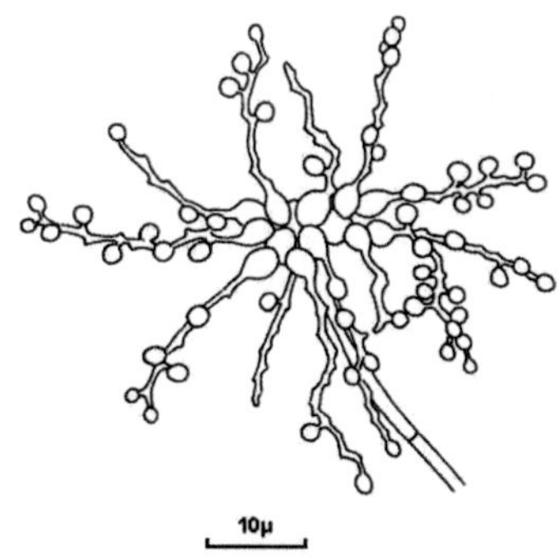

Fig. 19.4: Morphology of *Beauveria bassiana*

When the conidium comes into touch with the silkworm's body, it germinates in 8 to 10 hours under ideal temperature and humidity circumstances. In addition to releasing its germ tube during germination, the conidium also secretes chitinase, which helps the germ tube pierce the body wall and continue to multiply. Following its invasion of the larvae's blood, the conidium's germination tube matures into vegetative hyphae, which have short, round, or oval-shaped tips. Frequently, they separate and grow longer to create vegetative hyphae. Aerial hyphae containing many conidiophores are formed when vegetative hyphae emerge from the skin. One or two conidia are carried by the tiny branches that emerge from these conidiophores.

Symptoms

When an infection is first developing, there are no noticeable symptoms; but, as the illness progresses, wet spots start to show up on the skin. Larvae at this stage become sluggish and lose their appetite. The larvae's body becoming limp, their skin loses its elasticity, they cease moving, and ultimately, they perish. Vomiting and diarrhoea symptoms start to show up before death. The body is initially mushy after death, but it stiffens and becomes hard after 6 to 8 hours. The body's colour is pink at this point. This is because the secondary bacterium *Serratia marcescens* is multiplying. Woolly aerial hyphae emerge between the intersegmental membrane one to two days later. The entire body is then covered in white, powdery conidia, with the

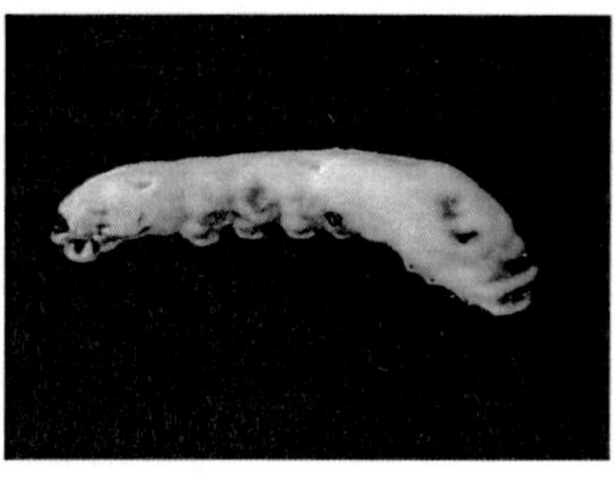

Fig. 19.5: White muscardine disease caused by *Beauveria bassiana*

exception of the chitinous areas of the head. When a pupal infection occurs, the pupae slow down their response to external stimuli. The abdomen wrinkles and the thorax shrinks. The hyphae in the air and conidia can enlarge to a third of their normal weight within the cocoons. These cocoons resemble dry shaken into cocoons. The moth stage is when the body is toughened and the wings effortlessly come off (**Fig.19.5**).

Prevention and control

Rooms, appliances and the surrounding area must be completely cleaned with a solution of 5% bleaching powder or 2% formalin before silkworm raising can begin. Avoiding low temperatures and high humidity during raising can help control this disease because these conditions are more conducive to fungal growth. Rearing beds should be kept as dry and thin as possible to prevent conidia from germinating and the fungus from spreading. Disinfecting and replacing trays, rearing sheets, cleaning nets, foam pads, and other items is necessary if the disease is discovered during rearing. Before being mummified, diseased worms should be carefully removed and placed in lime jars, burned, or deeply buried with a disinfectant applied. Likewise, the excrement and bed waste need to be disposed of appropriately. Garden pests that are mulberry-related should be managed since they are quickly infected with this fungus and subsequently serve as a major source of cross-contamination for silkworms.

Application of formalin chaff: The formalin solution concentration needed for this procedure varies based on the instar of the silkworm. The larval body and bed are treated with a mixture of burnt paddy husk and Formalin chaff. Formalin concentrations of 0.4% during I and II instars, 0.5% in III instar, 0.6% in IV instar and 0.8% in V instar are needed. The paddy husk is burned, charred or roasted in a pan without producing ash. The necessary strength of formalin is combined with the burned paddy husk in a volumetric ratio of 1:10 and the mixture is thoroughly mixed, depending on the stage of larvae development. Next, it is uniformly dispersed across the larvae. Then it is sprinkled evenly on the larvae and covered with a paraffin or double fold newspaper. After half an hour, the paper cover is removed and feed is given. Formalin chaff application should not be done when larvae are preparing for moult or under moult. Application of formalin chaff can be done before brushing on the newly hatched larvae and after each moult, half an hour before the resumption of feeding. The frequency of application of formalin chaff should be increased depending on the incidence of disease.

Application of Dithane M 45 (Zinc ion Manganese ethylene oxide bisdithio carbamate) or captan (N-Trichloromethyl Thio-4-Cyclohexane 1,2-Dicarboximide): These are the two widely accessible fungicides that are used to manage muscardine. In conjunction with levigated China clay or kaolin, these fungicides are applied at a concentration of 1% during I, II, and III instars and 2% during IV and V instars. The components are well combined, wrapped in a thin piece of cloth, and sprinkled on the freshly hatched larvae. Each moult is followed by a half-hour wait before feeding again. For instars I, II, and III, the required amount is 2-3 g per 0.1 sqm area; for instars IV and V it is 4-5 g. When the larvae are either in the process of moulting or have just begun to do so, dusting them is not advised.

19.3 Pests of Mulberry Silkworm and Their Control

In addition to being vulnerable to many diseases, *Bombyx mori* L., the silkworm, is also targeted by several pests. *Exorista sorbillans* Wiedemann, often known as the uzi fly, is a dipteran parasitoid that is the most dangerous insect-pest that attacks silkworms.

19.3.1 Major Pests

19.3.1.1 Uzi Fly

Exorista sorbillans Wiedemann (=*Exorista bombycis* = *Tricholyga bombycis* = *Tricholyga sorbillans* = *Tachina sorbillans*)

This parasitoid is a member of the family Tachinidae of the order Diptera, specifically the Tribe *Exoristini* within the sub-family Goniinae. In Bangladesh, China, Japan, India, South Korea, Thailand, and Vietnam, there have been reports of presence. In the tropical sericulture region, which includes Bangladesh, the southern portion of China, India, Thailand, and Vietnam, the fly's incidence is quite high. The percentage of damage varies between 10 and 30%.

Adults are blackish gray in colour. Male is longer in body length (about 12 mm) than female (about 10 mm). The head is triangular in shape. On the dorsal side of the thorax, there are four longitudinal black bands. The abdomen is conical. Life span of adult flies varies with sex and season. Males survive for about 10-18 days. Females live 2-3 days longer than the males. Survival period is least during summer months. Male can be distinguished from the female by the presence of external genitalia covered with brownish orange hairs on the ventral side of the abdominal tip. The female lays fertilized eggs all the way through her life. There are not many eggs laid at first. It rises gradually and peaks between the fourth and seventh day following adult eclosion. Macro type eggs have a creamy white colour. The length and width of the egg are 0.45 - 0.56 mm and 0.25 - 0.30 mm, respectively. The maggot burrows into the silkworm's body after hatching. Usually facing the silkworm body, the operculum allows the immature maggot to emerge from the eggshell. The newly hatched maggot enters the silkworm's body straight away. Maggots pass through three instars. In the first two instars, they develop just below the skin of the host body and in the final instar they leave this site and move into the body cavity. Pupation takes place in the darker area in and around the silkworm rearing house like corners, crevices, silkworm rearing beds, cracks of the rearing stands or trays, below the ant wells or in the superficial layer of the soil. Pupae are oblong in shape, somewhat oval anteriorly and round posteriorly. They are light reddish brown to dark reddish brown in cooler.

Type of damage and symptoms

Before they can reach the spinning stage, the silkworm larvae infected up to the early fifth instar perish. When an infestation occurs during the late fifth instar, the mature maggot emerges through the cocoon's perforation, making the cocoons unsuitable for mass reeling. A black scar on the area of skin where the maggot enters the host larvae's body indicates that the silkworms or pupae are infested. On the larvae's skin, tiny, creamy white oval eggs that are smaller than a pinhead are seen during the early stages of infestation (**Fig.19.6**).

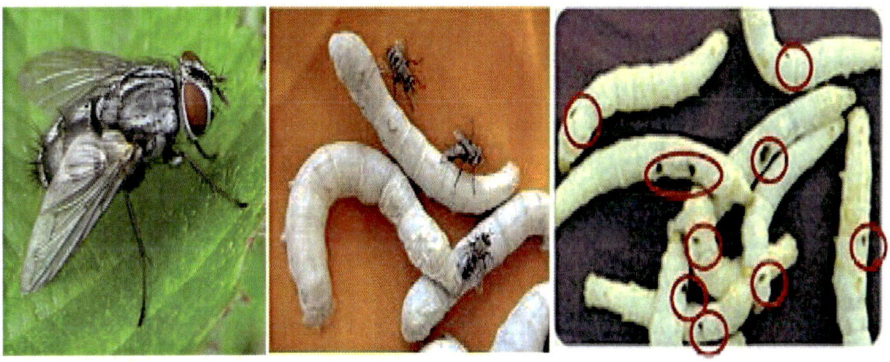

Fig. 19.6: Damage by uzi fly

Prevention and control

(i) **Physical**: Putting in wire mesh for the rearing room windows and doors or covering the rearing stands with mosquito net drapes to create a physical barrier. These techniques have been shown to be quite successful in limiting the amount of crop damage by keeping the female uzi fly from reaching the silkworms and laying eggs on them. It is important to take precautions to ensure that curtains are not disturbed and doors are not left open. By dusting levigated China clay on the silkworm's body during mounting, a physical barrier that stops the uzi fly from ovipositing can also be formed between the silkworm and fly. Doses of 3-4 g are advised for every 100 spinning larvae and square feet of the spinning tray.

(ii) **Chemical**: A commercial formulation uzicide containing 1% of Benzoic acid has been developed in India which kills the eggs of uzi fly when applied within 48h of egg laying. This treatment has to be repeated on alternate days. The dosage recommended is 7-8 ml/sq. ft. area.

19.3.2 Minor Pests

Mulberry silkworm is also attacked by several species of other insects like Coleopterans, *Alphitobius laevigatus* (Fab.), *Necrobie rufipes* (Degeer), *Tribolium castaeneum* (Hbst.) and many species of genera Dermestes; a dermapteran *Labia arachidis* (Yersin); and an acarid *Pediculoids ventricosus* (Newport). These pests are generally of minor economic importance in restricted localities.

19.3.2.1 Dermestid Beetles

This group of insects belong to the family Dermestidae of the order Coleoptera. They often attack pupae and adult silkworms in the grainages, and mostly cause extensive damage to the stored cocoons. They occur throughout the year.

19.3.2.1.1 D. cadverinus

Adults are oval elongate and dark brown in colour measuring about 1 cm in body length. The female beetle deposits its eggs in the crevices, eggs are oval, elongated about 2 mm in length and milky-white in colour. Eggs hatch in about 7 days. The spindle shaped grubs are reddish brown in colour. They moult 5-7 times in about 1-2 months and attains a length of about 1.5 cm. Body of the grubs are covered with hairs (**Fig.19.7**).

Fig. 19.7: Damage caused by dermestid beetles to the egg laying moths

Prevention and control

Rejected cocoons and dead eggs should not be kept in storage for an extended length of time. Periodic cleaning is necessary for both the cocoon storage rooms and the raising home. The grainage area needs to be cleaned both before and after the silk moth emerges. Grainage and wooden items kept in storage should be immersed in a 0.2% malathion solution for 2-3 minutes. To reuse the trays, they must be completely cleaned in water after 10 days and allowed to dry in the sun for two to three days. The insects are killed by hot air (50–60°C) being passed into the storage rooms and by keeping the humidity at 30% or lower. All stages of the beetle are killed by fumigating a dried cocoon storage room for three days with 0.5 g of methyl bromide per 3 m^2. Necessary precautions must be followed for using this chemical as fumigant.

19.3.2.2 Straw Itch Mite

Pediculoides ventricosus (Newport) (Family: Peduculoididae)

It has been reported in China, India and Japan. The adult female has an oval shape, but the male has a spindle shape. Head is triangular. There are four pairs of legs with tiny claws on the thorax and abdomen. It is an ovoviviparous mite. The nymph emerges from the eggs inside the female's body and eventually take on the size of miniature adults. About 100-150 young ones are produced by each adult female. The newly born mite has a body length of approximately 0.2 mm and a colour of light yellow. First the males and then the females are generated. After mating with a few females, each male perishes after a days or so.

Damage type and symptoms

This acarid attacks silkworm larvae, pupae, and adults. Attacks mostly target the juvenile larval and pupal stages. In order to get nourishment, the females tuck themselves into the delicate skin that lies between the host's larval and pupal segments. Moreover, a type of toxin found in their saliva eventually kills the victim. The silkworms vomit yellowish brown fluid, turn purple-brown in colour, and nearly stop eating after becoming infested. Defecation is challenging, as the anus is

Fig. 19.8: Straw itch mite

covered in beaded faeces. The attacked host's skin surface has several rough, irregular black patches. Attacked silkworms during their moulting are unable to shed their skin (**Fig.19.8**).

Prevention and control

On discovery of the attack by this acarid, the rearing rooms and trays should be replaced. The rearing appliances should be disinfected with steam. Cotton, wheat straw or rice straw should not be kept to dry near the rearing rooms or rearing appliances. Pre-rearing disinfection should be carried out with sulphur @ 30 g sulphur to a cubic foot or spraying of the rearing room and appliances with a 1 : 500 solution of an acaricide.

References

Handbook on pest and disease control of mulberry and silkworm: ESCAP team, Dr. K. Sengupta from the Central Sericultural Research and Training Institute of Mysore, India, ST/ ESCAP/888

Outcomes Assessment

Part A

Answer the following questions true or false. [1×5=5 Marks]

1. *Nosema bombycis* is caused by bacteria. [True/False]
2. Uzi fly is the pest of mulberry silkworm. [True/False]
3. Formalin is used as disinfectant for the rearing of silkworm. [True/False]
4. Dermestid beetles feed on the mulberry silkworm larva. [True/False]
5. Grasserie disease in mulberry silkworm is caused by NPV. [True/False]

Part B

Answer the following questions. [2×5=10 Marks]

1. Damage symptoms of flacherie disease.
2. Damage symptoms of grasserie disease.
3. Damage symptoms of dermestid beetles.
4. Control measures of straw mite.
5. Marks of identification of uzi fly.

Part C

Write a brief note on each of the following. [5×4=20 Marks]

1. Write short note on symptoms of protozoan disease
2. Write about the control measures of grasserie disease.
3. Write short note on symptoms caused by uzi fly.
4. Write about the symptoms caused by fungal disease.

Part D

Select the correct answer from the four choices given under each question. [1×5=5 Marks]

1. NPV is caused by
   - a. Nosema
   - *b. Borrelina*
   - *c. Exorista*
   - d. Straw mite
2. Grasserie diesase is also known as
   - a. jaundice
   - b. diarrhea
   - c. pepper
   - d. None of the above
3. Protozoan disease is caused by
   - a. *Exorista sorbillans*
   - b. *Nosema bombycis*
   - c. *Borellina* spp
   - d. *Tribolium castaneum*
4. Scientific name of uzi fly
   - a. *Tribolium castaneum*
   - *b. Nosema bombycis*
   - *c. Exorista sorbillans*
   - d. All of the above
5. *Beauveria bassiana* is known as
   - a. Green muscardine
   - b. White muscardine
   - c. Yellow muscardine
   - d. All of the above

Answer Key: 1. **b**, 2. **a**, 3. **b**, 4. **c**, 5. **b**.

20

Study of Reeling Equipment

20.1 Introduction

The technique of reeling several cocoon baves together to create a single thread is known as silk reeling. This is accomplished by winding the resulting thread onto a rapidly spinning reel after collectively unwinding filaments from a collection of cooked cocoons at one end in a heated water bath. There are three different ways to reel in raw silk: (1) directly on a standard-sized reel, (2) indirectly on small reels, and (3) by using a re-reeling machine to transfer the reeled silk from small reels onto standard-sized reels. An efficient reeler will hold onto the predefined number of cocoons per end in order to produce silk with a consistent denier Some techniques are used in reeling. Different reeling mechanisms are used for different purposes.

20.2 Various Silk Reeling Devices

20.2.1 Hand Spinning Wheel

Using a simple spinning method, two hands run the apparatus to remove silk thread from cocoons: one hand turns the wheel while the other feeds the cocoons. In one of the devices, the cocoons are cooked, while in the other, reeling thread is wound onto each wheel. A more tactile and artisanal method of silk reeling is available with hand spinning wheel reeling. A great degree of expertise and experience are needed to provide consistent outcomes and the appropriate silk characteristics (**Fig.20.1**).

Fig. 20.1: Hand spinning wheel

20.2.2 Charkha Type Reeling Machine

This is the traditional method of reeling silk in India, using the charkha equipment. Roughly half of the world's raw silk is produced in Charkha.

Reeling machinery requires very little initial investment because it is typically manufactured at home by a carpenter and a blacksmith using wooden materials, makes use of the direct reeling method, in which the cocoons are manually fed to the ends after the threads are reeled onto one end. Both cooking and brushing are done in the same basin. This machine makes a continuous silk thread, eliminating the need for re-reeling (**Fig.20.2**).

Fig. 20.2: Charkha type reeling machine

Advantages

Lower grade cocoons can also be retrieved by the charkha. Compared to other reeling machines, the cost of manufacturing is quite low, requiring only two persons for maintenance that requires very little labour.

Disadvantages

The silk that is spun using the charkha method is of low grade. Taking into consideration of the boiling and reeling processes, basin water becomes muddy and discolour, dulling the raw silk. Fuel costs increase when water needs to be heated and replaced on a regular basis. Furthermore, the reeler has to replenish the basin four or five times a day, which results in breakage and loose ends.

20.2.3 Cottage Basins or Domestic Basins

A common method of traditional silk reeling used in cottage or farm-based enterprises is the cottage basin. In contrast to commercial methods for reeling silk, this technique is used on a smaller size and requires a streamlined configuration using fundamental apparatus.

20.2.3.1 Domestic Basin

The domestic basin improves the quality of raw silk to compensate for the inadequacies of the charkha reeling system. One cooking basin and two reeling basins at a reasonable height make up the domestic basin unit. To cook and transfer cocoons, the two reeling basins are placed in front of the cooking basin. After cooking in a single pan, cocoons are moved to the reeling basin.

20.2.3.2 Cottage Basin

Fig. 20.3: CSTRI 8 ends improved cottage basin reeling machine

Cottage basins are commonly used in reeling and the cooking unit has three or four of them arranged in a row. Cooking does not interfere with reeling labour. The cocoons are cooked and reeled separately in a cottage basin to enhance the quality of the silk. There are reeling basins fastened to the reel bench. There will be six to ten ends per basin. A plan is made to fill each basin with hot water and empty the water of any debris that has accumulated from floss and sericin.

CSTRI has made the following changes to the existing cottage basin (**Fig.20.3**):

- Modification of thread passage
- Introduction by Jettebout
- Standard conventional rollers are swapped out with croissure pulleys.
- Swapping wooden little reels for plastic ones.
- Using ceramic buttons rather than conventional reeling buttons that have holes that are too small.
- Including a simple individual reel stop motion and a side brushing mechanism.
- It is a good idea to periodically remove any garbage that has stuck to the jetting booth.

Advantages

Compared to charkha silk, the quality of the manufactured silk is higher. The silk quality will be maintained since fresh water is used in the reeling basin. It can boost everyday productivity because it is powered.

20.2.4 Multi-End Reeling Machine

Slow motion reeling and thread manufacture on small reels at several ends per basin are the fundamental ideas of filature or multi-end reeling machines.

Fig. 20.4: Multi-end reeling machine

The multi ends machine eliminates the shortcomings of the seated type reeling machine by reducing the reeling pace. Thread is created on tiny reels by multi-end reeling devices that spin slowly. Reeling gently to break the thread results in less waste and easier re-reeling. It not only enhances mechanical silk output but also lessens worker strain. This reeling method has changed both design and functioning. The multi-end unit automatically cooks the cocoons. The reeling unit consists of two rows of basins with small overhand reels arranged in tandem. Every reeling machine has porcelain button thread guides with a tiny thread aperture. Because they have speed governors and are vibration-free, the machines last longer. The multi-end machine improves reeled silk's quality, production and waste. In every basin, there are about twenty times as many reeling threads. Because of the machine's slower speed, the silk strands it produces are of a much higher quality (**Fig. 20.4**).

Advantages

It is possible to produce superior quality silk. Less garbage is produced. A high production rate, less time-consuming. Designs are created to control the tension of the threads and the pace of the reels.

20.2.5 Advanced Silk Reeling Methods

Advanced silk reeling techniques have emerged as a result of technological advancement and the need for increased productivity and efficiency in the silk production process. In order to automate the procedure and enable faster and more dependable filament extraction, mechanical reeling machines have been designed. These devices use cutting-edge methods to unwind the silk threads from the cocoons, saving time and labour. Furthermore, chemical methods have been developed, which use certain chemicals to dissolve sericin and facilitate the extraction of silk threads. These chemical reeling techniques provide exact control over the parameters of the silk reeling process and can be tailored to produce desired silk properties (**Fig.20.5**).

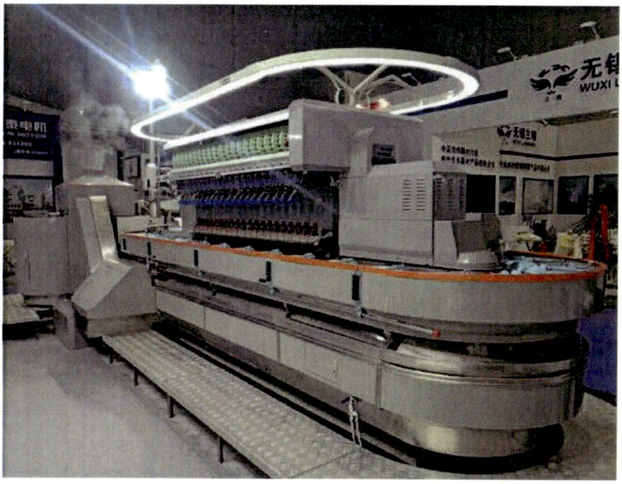

Fig. 20.5: Advanced silk reeling methods

20.2.6 Automatic Reeling Machine

In the past, reeling silk required labour-intensive hand removal of the silk thread from the cocoon by labourers. However, the procedure has been expedited and streamlined with the advent of automatic reeling machines. Cocoons can be mechanically fed into the reeling process thanks to a mechanism built inside the machine. Selecting and categorizing silk cocoons based on their kind and quality is the first stage. The weight, size and colour of the cocoons are all taken into account by a computerized system that maintains consistency in the spun silk yarn.

Following sorting, the cocoons are taken to the boiling section where they are boiled and then the silk filaments are unwound from the cocoons using a rotating reel. The automatic tension control systems of the reel ensure a consistent and even pulling force, resulting in the production of high-quality silk yarn. When the silk is being recovered for use or storage, it is wound into bobbins or spools. This is necessary to maintain the quality and purity of the silk thread. For bi-voltine cocoons automatic

Fig. 20.6: Automatic reeling machine

reeling machines are the best option. This apparatus requires very high-quality, uniformly sized cocoons with very little floss (**Fig.20.6**).

Advantages

In the event that a silk filament breaks during the reeling process, the machine may have mechanisms to reattach the damaged filaments without compromising the overall quality of the silk thread. An automatic reeling device will create superior silk. Automatic reeling devices are designed to remove silk from cocoons faster than with traditional methods. Since they can process more cocoons in less time, they raise productivity levels all around. They minimize the labour-intensive nature of silk reeling and help to maintain uniform quality standards.

Disadvantages

It is not possible to automatically reel in the multi-voltine cocoons.

20.2.7 Bobbin Silk Reeling Machine

A conventional automatic reeling machine was transformed into a brand-new type of bobbin silk reeling machine with cutting-edge winding and tension management technologies. Due to a lack of tension control mechanisms, automatic silk reeling machines—which are frequently employed in the traditional silk-making process to manufacture silk slices on small reels cannot satisfy the production of silk slices on bobbins. Additionally, silk slices cannot be prepared on bobbins using the winding devices found in automatic silk reeling machines. To prepare silk slices on bobbins, it is consequently required to design special bobbin silk reeling machines.

By combining a Feiyu 507 automatic reeling silk experiment machine with modern advanced winding and tension control technology of winders, created a new kind of bobbin silk reeling machine. The Feiyu 507 automatic silk reeling experiment machine's supplying cocoons, feeding ends, denier detection, twisting and other features and equipment were kept when it was converted into the new type of bobbin silk reeling machine.

The original collective steam winding mode was changed into a single-thread electric drying mode, and the centrally controlled winding mechanism for small reels was transformed into a single-thread controlled winding device for bobbins. An oiling device, an overfeed device; a tension adjustment device and an interrupt monitoring were fitted before the winding device (**Fig. 20.7**).

Advantages

The molecular make-up and the crystal structure of the raw silk were largely unaffected by the usage of the new kind of bobbin silk reeling machine. The traditional silk slices on bobbins had raw silk that was supple and ideal for weaving processing, similar to standard silk slices on bobbins. Cutting down on the number of steps required to prepare raw silk, such as re-reeling, finishing and steeping by using the new type of bobbin silk reeling machine to prepare silk slices on bobbins directly, can also lower production costs, increase productivity, use less energy and protect the environment.

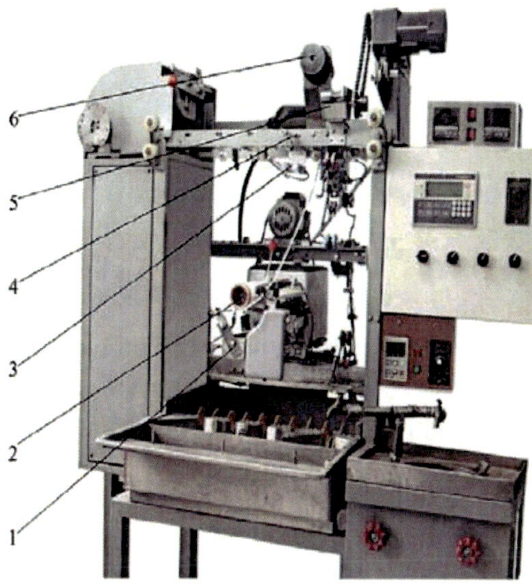

Fig. 20.7: Bobbin silk reeling machine

20.2.8 CSTRI Bulky Raw Silk Reeling Machine

A component of the brushing mechanism of the automatic silk reeling machine served as the inspiration for the development of the bulky raw silk reeling machine. Typically, 8 to 10 silk filaments were combined on a vertical axis in a reeling machine to generate mulberry raw silk. The CSTRI differential traverse mechanism is used in the bulky raw silk reeling machine to collect the cocoon filaments on the horizontal axis. The mechanism can move in two directions: perpendicular to its axis and on it. The rotary traverse mechanism rotates by encircling the cocoon filaments while simultaneously moving them forward, causing the filaments to be arranged in a zigzag pattern. The finished yarn has a web-like structure and is thicker than raw silk made from an equal number of cocoons in conventional reeling machines. It is created using the bulky raw silk reeling machine (**Fig.20.8**).

Fig. 20.8: CSTRI bulky raw silk reeling machine

20.3 Conclusion

Though large-scale silk production has mostly moved away from manual reeling due to the introduction of new machinery, silk reeling techniques are still practiced today in certain parts of the world. Reeling silk is still a significant traditional craft that helps to preserve the art of producing silk and increases the market for high-quality silk fabrics. These tools have evolved into essential tools for modern sericulture practices, enabling silk producers to better meet consumer demand. Exciting prospects lie ahead for silk reeling processes. Expectations include improvements in digital technology, silkworm genetic development, sustainable methods, personalization and integration with sustainable fashion. These developments seek to improve efficiency, quality and sustainability while preserving the rich history and cultural value of silk production.

References

Babu KM. 2012. Silk production and the future of natural silk manufacture. In Handbook of Natural Fibres: Woodhead Publishing

Gowrisankar R, Dasari S and Anas M. 2023. Silk Reeling Techniques: Exploring Traditional and Advanced Methods. Journal of Experimental Agriculture International. 45 (9): 91-99

Luo HL, Fu YQ and Liu K. 2020. Structural design of automatic silk reeling machine for direct silk winding. Journal of Textile Research. 41(8):115–120

Outcomes Assessment

Part A

Answer the following questions true or false. [1×5=5 Marks]

1. Cottage basins are commonly used in reeling. [True/False]
2. Modern method of reeling silk in India using the charkha machine. [True/False]
3. The silk that is spun using the charkha method is of low grade. [True/False]
4. Automatic reeling machines are the best for bi-voltine cocoons. [True/False]
5. Multi-End Reeling Machine produces inferior quality silk. [True/False]

Part B

Answer the following questions. [2 x 4 = 8 Marks]

1. Write about the advantages of bobbin silk reeling machine.
2. Write about cottage basins.
3. Write about charkha reeling.
4. Disadvantages of automatic reeling machine.

Part C

Write a brief note on each of the following. [5×4=20 Marks]

1. Write short note on advantages of charkha reeling
2. Write in brief about automatic reeling machine.
3. Write in brief about CSTRI Bulky Raw Silk Reeling Machine.
4. Write about the advantages automatic reeling machine.

Part D

Select the correct answer from the four choices given under each question. [1×5=5 Marks]

1. Charkha machine produces

   a. Low quality of silk
   b. High quality of silk
   c. Both a and b
   d. None of the above

2. Automatic reeling machines uses
   - a. Univoltine cocoons
   - b. Bivoltine cocoons
   - c. Multivoltine cocoons
   - d. All of the above
3. Conventional automatic reeling machine was transformed into
   - a. CSTRI machine
   - b. Bobbin silk reeling machine
   - c. Multi end reeling machine
   - d. None of the above
4. Multi-End Reeling Machine produces
   - a. Superior quality
   - b. Inferior quality
   - c. Both a and b
   - d. None of the above
5. A common method of traditional silk reeling used in cottage or farm-based enterprises is known as
   - a. Multi end reeling machine
   - b. Automated reeling machine
   - c. Cottage basin
   - d. All of the above

Answer Key: 1. **a**, 2. **b**, 3. **b**, 4. **a**, 5. **c.**

Index